Also by Charles Pellegrino

Time Gate
Her Name, *Titanic*
Flying to Valhalla (*fiction*)

with Jesse Stoff
Darwin's Universe
Chronic Fatigue Syndrome

with Josh Stoff
Chariots for Apollo

*with James Powell and
Isaac Asimov*
Interstellar Travel and
Communication

UNEARTHING ATLANTIS

UNEARTHING ATLANTIS

An Archaeological Odyssey

CHARLES PELLEGRINO

RANDOM HOUSE

NEW YORK

Copyright © 1991 by Charles R. Pellegrino
Foreword copyright © 1991 by Arthur C. Clarke
Maps copyright © 1991 by Anita Karl and Jim Kemp

All rights reserved under International and Pan-American Copyright Conventions.
Published in the United States by Random House, Inc., New York and
simultaneously in Canada by Random House of Canada Limited, Toronto.

Grateful acknowledgment is made to the following to
reprint previously published material:
HARCOURT BRACE JOVANOVICH, INC., and FABER AND FABER LIMITED:
Excerpt from "Burnt Norton" in *Four Quartets* by T. S. Eliot.
Copyright 1943 by T. S. Eliot and copyright renewed 1971 by
Esme Valerie Eliot. Rights throughout the world excluding
the United States are controlled by Faber and Faber Limited.
Reprinted by permission of Harcourt Brace Jovanovich, Inc.,
and Faber and Faber Limited.

CARL SAGAN: Excerpt from a speech given at Carnegie Hall, May 1984. Copyright © 1984
by Carl Sagan. Reprinted by permission of the author % Scott Meredith Literary Agency,
Inc.
WARNER/CHAPPELL MUSIC, INC. AND E'G MUSIC, INC.: Excerpt from "Once in a Lifetime"
by David Byrne, Chris Frantz, Tina Weymouth, Jerry Harrison, and Brian Eno. Copyright
© 1981 by Bleu Disque Music Co. Inc., Index Music, E'G Music Ltd., and Warner Bros.
Music Ltd. All rights reserved. Used by permission of Warner/Chappell Music, Inc., and
E'G Music, Inc.

Library of Congress Cataloging-in-Publication Data
Pellegrino, Charles R.
Unearthing Atlantis: an archaeological odyssey/Charles R.
Pellegrino.
p.   cm.
Includes bibliographical references.
ISBN 0-394-57550-4
1. Thera Island (Greece)—Antiquities.   2. Minoans—Greece—Thera
Island.   3. Atlantis.   I. Title.
DF221.T38P45   1990
939'.15—dc20   89-43543

Manufactured in the United States of America
98765432
First Edition

ILLUSTRATIONS BY C. R. PELLEGRINO
BOOK DESIGN BY JO ANNE METSCH

To five who believed in a "retarded" boy who could not read, but thought he would: Mom, Dad, Adelle Dobie, Barbara and Dennis Harris. Thank you for not believing in test scores, for not browbeating and for believing instead that you could bring the boy out by encouraging his love of science. Would that every school had five teachers like you.

Perhaps there is no more poignant theme than *if only . . .*
—Gregory Benford,
What Might Have Been

FOREWORD

Atlantis! There can be no name in the languages of the Western world that evokes more feelings of wonder, mystery—and irreparable loss. The myth created by Plato more than two millenia ago still has power over our age: even as I write, a space shuttle bearing that magic name is being prepared for launch at Cape Kennedy.

But was it only myth? Down the centuries, legions of amateur and professional scholars, covering the whole spectrum from sober historians to certifiable lunatics, have believed otherwise. They were correct—though not in the way that most of them imagined.

Of all the millions of words about Atlantis that have come down from antiquity, the most chilling are these: *"The Atlanteans never dreamed."* Should we envy them—or pity them?

I first encountered Atlantis via the lunatic fringe, thanks to a friendly neighbor with an interest in the occult. To quote from my "science-fictional autobiography" *Astounding Days:*

> Mrs. Kille stimulated my curiosity by lending me such books as Ignatius Donnelly's 1882 masterpiece of spurious scholarship, *Atlantis: The Antediluvian World*—the veracity of which I did not doubt for a moment, never imagining at that tender age that printed books could possibly contain anything but the truth.
>
> Donnelly should be the patron saint of the peddlers of UFO/parapsychology mind rot; not only did he claim that all the ancient civilizations were descended from Atlantis, but in his spare time (he was also a Philadelphia lawyer, lieutenant governor of Minnesota—and

twice a congressman) he "proved" that Bacon wrote Shakespeare, Marlowe—and Montaigne's essays!

That ten-year-old bookworm gulping down Donnelly's splendid nonsense certainly never imagined that half a century later he would be walking in the streets of the *real* Atlantis—without even knowing it! The tale which Charlie Pellegrino unfolds is one of the great archaeological detective stories of all time, and some of the most exciting chapters are yet to come.

At this point I feel I owe an apology to the several hundred other writers who over the last few years have received my notorious "Drop Dead" form letter saying that I *never* write prefaces, or even blurbs. (Four went out in a single mailing last week. . . .) Well, there have to be exceptions to every rule; and when writing *The Ghost from the Grand Banks* I was greatly indebted to Charlie for his advice, and especially for his book *Her Name, Titanic.*

I would like to end with my own tribute—from *The Songs of Distant Earth*—to the immortal myth, in the words I have left for some musician more than a thousand years hence—when Earth itself faces total destruction:

> When I wrote "Lamentation for Atlantis" I had no specific images in mind; I was concerned only with emotional reactions, not explicit scenes. I wanted the music to convey a sense of mystery, of sadness— of overwhelming loss. I was not trying to paint a sound portrait of ruined cities full of fish. Yet now something strange happens, whenever I hear the *Lento lugubre*—it's as if I'm seeing something that really exists.
>
> I'm standing in a great city square almost as large as St. Mark's or St. Peter's. All around are half-ruined buildings, like Greek temples, and overturned statues draped in seaweed, green fronds waving slowly back and forth. Everything is partly covered by a thick layer of silt. . . .
>
> I know, of course, that Plato's Atlantis never really existed. And for that very reason, it can never die. It will always be an ideal—a

dream of perfection—a goal to inspire men for all ages to come. . . .

Now, the "Lamentation" exists quite apart from me; it has taken on a life of its own. Even when Earth is gone, it will be speeding out toward the Andromeda galaxy, driven by fifty thousand megawatts from the deep space transmitter in Tsiolkovski crater.

Someday, centuries or millenia hence, it will be captured—and understood.

—Arthur C. Clarke
Colombo, Sri Lanka
April 1990

ACKNOWLEDGMENTS

I wish to acknowledge warmly the townspeople of Akrotiri, who not only made my stay there very pleasant, but who, along with various archaeologists and other scientists, including Nanno Marinatos (to whom I am most indebted), shared their recollections of life at the excavation during the past two decades. Without their contribution this would be a quite different book. These recollections (combined with excerpts from Spyridon Marinatos's published journals) are the basis for historic scenes at which I could not possibly have been present, and which no one would otherwise have known about. I have in a few instances drawn upon this material to conceive scenes involving Professor Marinatos which, while they must in a strict sense be considered fictional, are true to his thoughts, philosophies and published works.

For additional background on the life of Spyridon Marinatos, particularly during his first three years at the Thera excavation, I acknowledge (and highly recommend) J. V. Luce's 1969 book, *Lost Atlantis: New Light on an Old Legend*. I have relied throughout (as does Luce) on Benjamin Jowett's translations of Plato's *Timaeus* and *Critias*, which are reproduced in their entirety in Luce's book.

My thanks to Haraldur Sigurdsson, Jean Francheteau and Roger Hekinian, my shipmates on Expedition *Argo*—RISE, who shared with me a genuine love of volcanoes.

I thank the four Fourths for touring me through Love Canal—which is, next to the *Titanic*, the most haunting and haunted place I have ever known.

My wife Gloria accompanies me on my expeditions, and during

our visits to the island she raised many human elements of the Thera story that might have remained hidden had her questions gone unasked.

Many others contributed to this book during wide-ranging conversations over a four-year period. Those whose comments or insights were especially valuable include Jim Powell (Brookhaven National Laboratory), Arthur C. Clarke (University of Moratuwa, Sri Lanka), Father Mervyn Fernando (Subodhi, Institute for Integral Education, Sri Lanka), Father Robert A. McGuire (Charismatic House of Prayer, New York), Stephen Jay Gould (Harvard University), *the* Don Peterson (Natural Resource), Isaac Asimov (National Treasure), Walter Lord (National Treasure and all-around nice guy), Bill Schutt (National Disaster), Alan Branch (Wounded Knee), Ed Bishop (RPI), Steve Schoep (RIP), Jean Gonzalez (J-J Books), Mitch Cotter (AAAS), Daniel Stanley (Smithsonian Institution), Susan Limbrex (University of Birmingham, Great Britain), Jeffrey Soles (University of North Carolina), Carl Sagan (Cornell University), Lawrence Soderblom (NASA/JPL) and Mrs. Spyridon Marinatos.

I thank my agent, Russ Galen, and my editor, Rebecca Saletan, for getting this thing off the ground and providing important editorial insights throughout. Thanks also to Anne Sweeney for "flight control," to Patrick A. Lyons (Smith and Laguercia, P.C.) for "crash avoidance" and Harold Evans (R.H.) for "guidance systems."

I thank Tom Dettweiter, Cathy Scheer, Shelley Lauzon, Marge Stern, Terry Nielson, Ed Bland, Barry Walden, Larry Schumaker and Kirk McGeorge for their hospitality during my stays at Woods Hole, and for the wealth of information they provided about the workings of *Argo*. I am grateful to everyone who sailed with me on the the *Melville* and the *Atlantis II*, particularly Robert Ballard, Chet Ballard, Earl Young, "Skip," Martin Bowen, Emile Bergeron, Cindy van Dover, Chris von Alt, Jean Louis Cheninée, Robert Aller, Per Hall, Peter Rude, Reuben Baker, Ralph Hollis, David Sanders, John Salzig, Paul Tibbetts, Will Sellers and, of course,

Acknowledgments

Argus, Argo and *Alvin*. A true picture of what the seamounts were, and how and when they formed required extensive exploration, and for sheer simplicity, the ascent of Larry Schumaker and Robert Ballard up the side of Atlantis Seamount in Chapter 2 is a composite of several *Alvin* dives.

I wish to note that none of the people I have acknowledged and none of the scientists and archaeologists described herein are to be held accountable for any heresies that might have squeaked through to publication nor are the theories and opinions expressed in this book necessarily endorsed by any of them. Please note also that I answer all of my mail, and I invite discussion and criticism for the updating of future editions.

CONTENTS

Foreword by Arthur C. Clarke   xi

Acknowledgments   xv

Introduction   3

I. THE CELLARS OF TIME   11

  1. A Shriek in the Night   13

  2. Atlas Adrift   29

  3. A Killing Wind   53

  4. Bible Stories for Archaeologists   85

  5. Mediterranean Genesis   93

  6. The Three-Pound Time Machine   145

II. RESURRECTION   161

  7. Descent into Darkness   163

  8. Apprentice   175

  9. A Room Full of Lilies   191

10. The Edith Russell Syndrome   209

11. Queen Hatshepsut:

    The Difficulties of Dating Older Women   217

12. The Inheritors of Atlantis   249

III. ICARUS   271

13. Phaëthon Rising   273

14. On Probability and Possibility   285

Afterword: Where Are They Today?   297

Selected Bibliography   303

Index   309

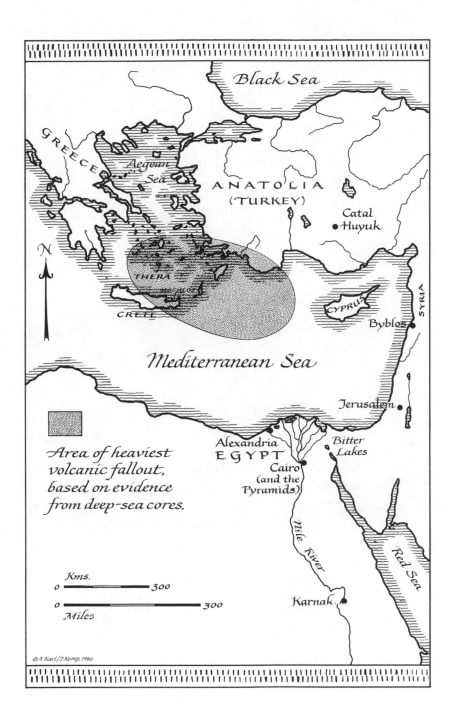

Black Sea

GREECE

Aegean
Sea

ANATOLIA
(TURKEY)

Catal
Huyuk

THERA

CRETE

CYPRUS

SYRIA

Byblos

Mediterranean Sea

Jerusalem

*Area of heaviest
volcanic fallout,
based on evidence
from deep-sea cores.*

Alexandria
EGYPT

Bitter
Lakes

Cairo
(and the
Pyramids)

Nile River

Red Sea

Kms.

0 ———— 300

0 ———— 300

Miles

Karnak

© A.Karl/J.Kemp, 1990

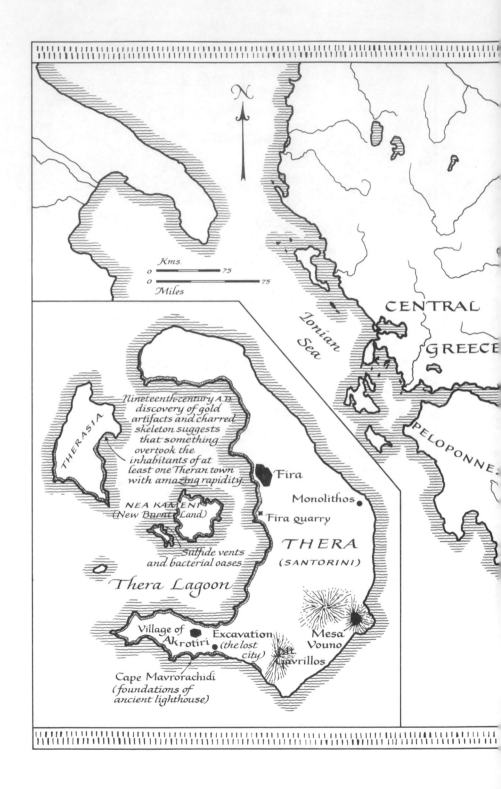

N

Kms.
0 ——————— 75
0 ——————— 75
Miles

Ionian
Sea

CENTRAL

GREECE

PELOPONNE

THERASIA

Nineteenth-century A.D.
discovery of gold
artifacts and charred
skeleton suggests
that something
overtook the
inhabitants of at
least one Theran town
with amazing rapidity

Fira

Monolithos

NEA KAMENI
(New Burnt Land)

Fira quarry

THERA
(SANTORINI)

Sulfide vents
and bacterial oases

Thera Lagoon

Village of
Akrotiri

Excavation
(the lost
city)

Mesa
Vouno

Gavrillos

Cape Mavrorachidi
(foundations of
ancient lighthouse)

Troy

Aegean Sea

LESBOS

SKYROS

ANATOLIA

(TURKEY)

ATTICA
Athens

ANDROS

A.Irini
KEA

TENOS

IKARIA

SAMOS

Turkish
tsunami
funnel

SYROS

MYKONOS

PAROS

NAXOS

KALYMNOS

KOS

Greek
tsunami
funnel

SIPHNOS

AMORGOS

MELOS

ASTYPALAEA

Trianda

(see inset)

THERA

RHODES

KARPATHOS

Harbor of
Herakleion

MOCHLOS

CRETE

Knossos

Zakro

Phaistos

© A·Karl/J·Kemp. 1990

UNEARTHING
ATLANTIS

The time will come when diligent research over long periods will bring to light things which now lie hidden. . . . Many discoveries are reserved for ages still to come, when memories of us will have been effaced. Our universe is a sorry little affair unless it has something for every age to investigate. Nature does not reveal her mysteries once and for all.

—Seneca, *Natural Questions*, Book 7, first century A.D.

INTRODUCTION

There is a place where a stone god rose up from the sea and a mountain opened up in the sky; and as the world heaved and compressed, there occurred sights and sounds of which today's foremost nuclear weapons designers scarcely dream.

I know of a world that saw waves as high as skyscrapers.

I know of a cloud that spread globular and huge, and where it touched the earth and the sea, it converted men into gas.

I know of a rend in the earth more than seven miles wide. It is flooded with water now, and if you were to stand in a boat near its center, your mind would resist coming to terms with the size of it, would try to reject the idea that such a thing could be carved out in a single day and night.

I know of a city near the rend, nearly four thousand years old and buried, in places, under more than two hundred feet of ash. Its storefronts and apartments are perfectly preserved, and they would not look out of place on a present-day street. The homes had running water and bathtubs and flush toilets; and it seems possible that bedrooms were heated in the wintertime by steam piped in from volcanic vents, as it traveled on to rooftop cisterns, where it was condensed for bathwater. If any were caught sleeping in those warm rooms when the volcano awoke, they were fortunate, in a fashion—more fortunate than the hundreds of thousands of their fellow Atlanteans on the shores of Crete, Naxos and Egypt. For they, at least, died without understanding that they were the last of the true Minoans, that their entire civilization had just been set upon the wane.

Atlantis, some will argue, was a large, mountainous continent somewhere in the Atlantic Ocean. But the Atlantis in our dreams and our myths was almost certainly the lost civilization of Minoan Crete. Ninety-five percent of what we know about this naval empire (known to the Egyptians as "the Keftiu") is being excavated from Thera, an island offshore of Crete, where a city whose name no one remembers was buried with all of its contents intact. During the summer of 1650 B.C., the city's tall, white buildings still gleamed under the clear, hot sky, and in the words of Plato, there grew on that island every variety of plant that is pleasant to the eye and good for food. In the center of the island there stood a volcanic peak. Its long, quiet sleep had lasted for tens of thousands of years, and might have lasted for tens of thousands more. But not forever. Nothing lasts forever.

The discovery of the buried city and the civilization that built it constitutes the most important archaeological find since Howard Carter and Lord Carnarvon entered the tomb of Tutankhamen. But King Tut's burial chamber was a relatively small thing. It could easily have been accommodated within one of the houses beneath Thera's vineyards, where entire streets are now being brought to light, along with frescoes showing us who these vanished people were and how they lived, still breathtakingly fresh more than 3,650 years after they were painted. The excavation has been progressing now for twenty-three years and has begun to look like an open pit mine, except that there is a city instead of coal on its floor, and this mine is being roofed over as the operation expands. But it progresses by inches, and so far has exposed less than a block of a buried city whose diameter is nearly a mile. Yet every building yields art treasures to rival anything since produced by the hand of man; and it's anyone's guess what lies just a few feet away behind the next wall of ash. All we do know is that hundreds of buildings remain buried and unexplored—enough of them, it seems, to keep archaeologists busy through the year 2300.

The city is a treasure chest whose lid has been pried open, ever so gently, to produce a very narrow crack through which we can see only glimmers of light. As one comes closer and actually begins touching fallen walls, and examining housewares crushed beneath them, the view grows wider and wider until finally through this same narrow crack we are looking out from the chest, out from Thera itself, at Cretan and Egyptian ruins, and beyond them to the high seas and deep under them.

My original intention was simply to bring the reader along to Thera, to the dig that is now giving us the first look we have ever had of the strangest, most romantic ancient world of all. Focusing on a single excavation on a far-flung Greek island, I'd planned to let the reader look through the eyes of paleontologists and archaeologists at rooms full of household items, and to show how, somewhat like detectives trying to reconstruct the scene of a murder, we can look at assemblages of objects and reconstruct the last days of a lost civilization.

But prehistoric cities do not lend themselves to tunnel vision. The island of Thera did not exist in a vacuum. On a lonely plain in the Wiltshire countryside, someone built tombs in the shadow of Stonehenge, and filled them with beads imported from Egypt around 1650 B.C. Either the Egyptians or their Minoan neighbors sailed to Britain, bringing the beads with them. We are often surprised by such discoveries, but we should not be. Ancient people simply traveled farther, and knew more, than we normally give them credit for. We are perhaps blinded by a recently ended millennium of dark ages.

Even trying to come up with a precise date for the end of Thera brings us far from Thera. We must look at mud samples from the deep sea. We must study volcanoes two miles down on the bed of the Atlantic. We must explore Irish peat bogs and the Greenland ice cap, learn everything we can about pine trees in California and, of course, study Cretan and Egyptian pottery styles. From these things, and more, we can say with reasonable certainty that a squall was blowing west to east across Thera when all life was suddenly extinguished there in the autumn of 1628 B.C. Unexpectedly, the excavation leads us to the Nile delta, Moses, the Exodus, and a chilling paradox that my Jesuit friends are all to happy to bring back to the Vatican.

For me, all of this began during the winters of 1985 and 1986 when, over cups of tea after scientific meetings, the name of Thera came up repeatedly in conversation. A number of colleagues, including the explorer Robert Ballard, had said, "You've got to go out there. If you've got any curiosity at all about archaeology, you've got to. You can probably catch a plane tomorrow. Better still, you might catch one tonight. Run, don't walk, and see if you don't respond to the site the way we did."

Two years would pass before I booked the flight, largely because I already had commitments to two ships, their names *Titanic* and *Valkyrie*. As it turns out, Thera was worth the wait. Beyond the excavation, the island itself was one of the most scenic places I'd encountered anywhere on earth; and during the summer and autumn of 1988, I fell in love with Thera.

At the outset, I could never have looked ahead and seen that Thera would lead me to heated discussions about Moses and statistical theology held, of all places, in the wilderness of Sri Lanka with a Jesuit priest and Arthur C. Clarke. I'd planned to live for a while on the island, piecing together clues accumulated during more than two decades of excavating and sending my editor letters sharing my understanding of what transpired in a nameless city between the years 1650 and 1628 B.C. I would add to this the story of Thera's chief archaeologists, Spyridon Marinatos and

Christos Doumas, who together threw open the doors to a world that lay hidden beneath the earth.

"Just write us letters," my editor had said. "Write as if you were writing to an intelligent friend. Tell us about Thera's last days. Tell us about Marinatos and Doumas. Tell it long."

That's how this book got started; but almost from my first day on the island, as I walked about picking up little chips of rock, I saw in them pieces of stories that altered the whole project. High above the excavation stands the limestone peak of Mesa Vouno. It is covered with Egyptian and Roman ruins, and as I kicked at the chalky earth, I dislodged a tiny clam identical in appearance to the ones commonly served up in the restaurants overlooking Thera Lagoon. But this clam had been off the menu for a very long time. Most of the original shell had been dissolved and replaced by stone untold millions of years ago. The fossil told me that Mesa Vouno had once lain on the bottom of the Aegean, though it now towered more than a thousand feet above sea level. I began to see the buried city and the island that enclosed it in a rather larger dimension than even the archaeologists imagine. I remember sitting on a fallen Egyptian pillar, looking south from Thera's highest point, looking toward the Nile. A sedimentologist had recently told me of a thin horizon of volcanic ash he'd just discovered between layers of Egyptian soil dating back to the Eighteenth Dynasty. Perhaps this, too, was part of Thera's story. What part, I could only guess at, but I would soon find out.

What follows is a detective story on a scale that sometimes makes me think Agatha Christie missed a most delightful calling. She should have been an archaeologist. There can be no doubt that she would have reveled in the Thera-Atlantis mystery.* Like every good whodunit, the search for clues to life and death on Thera requires that we move around a lot, often to places quite

---

* Though not an archaeologist herself, Agatha Christie was married to a famous archaeologist—Sir Max Mallowan, whose work probably inspired such books as *Death on the Nile*.

distant from the death scene, places whose connection to it may not at first be apparent. I have looked at the bottoms of oceans, at New Zealand's Kauri forests and at the abandoned streets of Love Canal. In Chapter Three I investigate the particularly gruesome effects of the Mount Pelée and Mount St. Helens explosions. Eyewitness accounts (which at St. Helens were augmented by data from scientific instruments) provide important insights into volcanic manifestations that, on Thera, Crete and in the Nile delta, have left only faint geologic signatures. The imagination is rarely more forcefully awakened than when one applies what can be learned from the St. Pierre and St. Helens death clouds to a reconstruction of the last days of Thera, and realizes with a start that, as horrifying as those two volcanic catastrophes were, Thera was thousands of times worse.

Tracking the origin of the Atlantis legend leads through volcanology, paleontology, classical literature, world history, the Bible and even into outer space. Even the soil that encloses the buried city precludes studying Thera in isolation. In the ash above we find Greek, then Egyptian and Roman artifacts, and higher up, on the surface, debris from the Christian era. Immediately below the city are the ruins of another city. Below them, fossil clam beds and, lower still, dinosaur teeth. So my reports on the buried city may range a little deeper than you anticipated and you may begin to feel as if you have come unstuck in time. Don't worry. If I've accomplished some modest measure of what I set out to do, that's exactly how you're supposed to feel. As a paleontologist who sometimes surfaces into the shallows of archaeological time, that's how I normally feel. So welcome to my world.

Charles Pellegrino
Rockville Centre, New York
April 1990

THE
CELLARS
OF
TIME

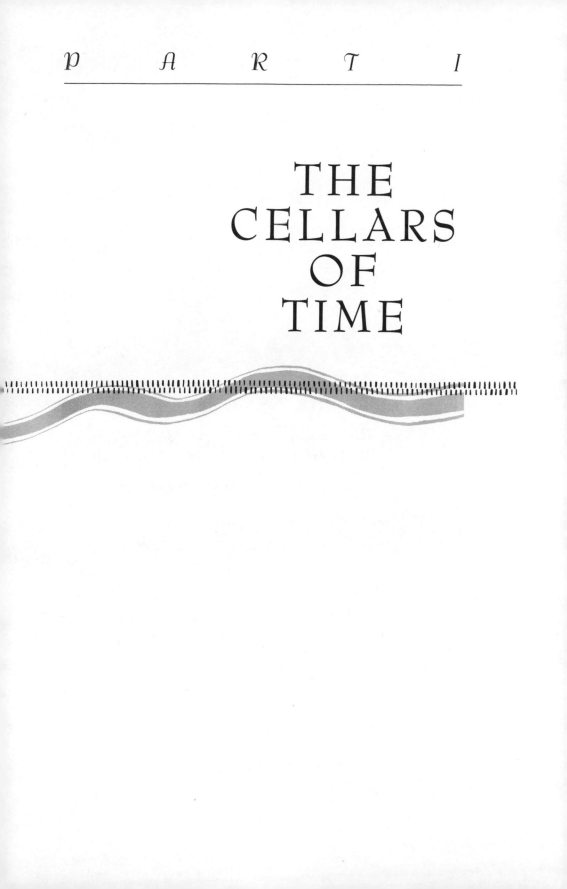

Then listen, Socrates, to a tale which,
though strange, is certainly true.

—Plato, *Timaeus*

A
SHRIEK
IN
THE
NIGHT

AUTUMN, 1628 B.C.

Down there on earth, the course of history was shifting, shifting like sand from infinite possibilities to the shadow of a dream.

Even from a vantage point on the lunar highlands, you could have seen it without the aid of telescopes or binoculars. A black spot had appeared on the Aegean Sea, perfectly round, vast and cloudlike. It swelled on the planet below, burst up through the atmosphere and started to rain fire down upon Melos, Naxos and Crete. At its center, the island of Thera was not. Walls of water hovered around a hole more than a mile deep. The air was hotter than live steam, hotter than molten lead, hotter than iron emerging white from a furnace.

The shadow on the earth grew with extraordinary speed, slashing down ships as it sluiced into the Mediterranean. White clouds formed and vanished on its edges. In less than an hour it widened to two hundred miles, and then to four hundred miles. It rolled over Turkey and Egypt, turned day into starless night, shed some

Within hours of the Theran upheaval of 1628 B.C., death rolled into southern Turkey on the tongue of a tsunami. Two peninsulas jutting into the Aegean Sea confined the wave as if between the prongs of a mighty tuning fork, building it higher and higher and ultimately funneling it thirty miles inland. To penetrate so far, it had to be eight hundred feet tall when it hit the shore. The skyline of lower Manhattan is a familiar and instructive guage for the magnitude of the Theran tsunami.

of its heat and ash and velocity, then stumbled eastward. The stain thinned over Syria and Iran, pulled itself apart into long streamers, thinned still more and spread across Asia like ink poured into a saucer of water—except that this ink stain was laced with sheets of brilliant blue lightning.

Passing now into legend and mythology, lost almost beyond recall in the stream of time, was the most progressive civilization the world had ever seen.

AUTUMN, A.D. 1988

When I sailed into the central lagoon of Thera (or Kalliste, as the ancients called it) six days ago, the Aegean was as calm as quarry water. Four miles away, in every direction, vertical cliffs rose more than a thousand feet out of the sea. I knew that four thousand years ago those same cliffs had been the mere foothills of a mountain that had stood nearly a mile over my head. It is easy to talk about cubic miles of rock vanishing from a place, but one must actually sail into the crater to understand the size of it. I've never received such an impression, not standing on Tasman Glacier, or in the shadow of Mount Cook, or even exploring parts of the forty-thousand-mile seam that runs along the ocean basins. I expected them to be big. They were built up over thousands, even millions of years, inches at a time. It's easy to understand small changes producing large results, given a sufficiently large amount of time. But my perception of time and space bends at Thera, because the hole in the earth was dug out in a geological nanosecond, during an event so violent as to be both fascinating and beautiful, especially when viewed, in the mind's eye, from a calm sea, with nearly a quarter mile of water below and the cliffs, the edges of the explosion, appearing distant on the horizon.

Until you visit this place, a description of the eruption amounts to barely more than numbers—insofar as we know them—rolling off the tongue. Near 1600 B.C., a city—probably two or three cities—died here. The volcano kicked up block after block of multi-storied apartment houses, bakeshops and storefronts, pulverized them and sank them out at sea, or hurled their dust at the stratosphere. One city, or a large portion of it, had already been buried under a blizzard of volcanic ash, on a part of the island that was not cracked open and flung about or heaved into the Aegean. The ash acted as a cushion, protecting the ruins from an explosive force exceeding 150 hydrogen bombs detonating at once. Fifty cubic miles of rock became as a vapor in the heavens, and death rolled into Turkey on the tongue of a tsunami. The wave was funneled thirty miles inland. To penetrate so far, it had to be eight hundred feet high. By today's standards, the wave would reach almost to the top of Manhattan's Chrysler Building.

Thera, which means "fear," is a flooded crater 150 miles southeast of Athens. To the south, the island of Crete stands between it and Africa, and bisects the Mediterranean and Aegean seas. Thirty-six hundred years ago Thera was a round island with a peak in its center, and fast-flowing streams, and a green canopy of papyrus and palms—if paintings within its buried buildings are to be believed. The island was called Kalliste then—"the most beautiful"—and there was a city on it, very advanced for its time. The Kallistens had close connections with Crete, which was at the height of its powers. We call these people Minoans, after Minos, the king whose reign dated back to a time that was already misty and vague then, and laden with legends like that of the Minotaur and the Labyrinth. At the time of the Thera upheaval, their civilization had already existed for fifteen hundred years. They'd built the world's first navy, traded with Egypt, and gained control of the eastern Mediterranean and much of the Greek mainland. The city on Thera's south shore lacked defensive, fortlike walls. The sea was its wall, and the navy defended the barrier. Thera shared in Minoan power, and grew to become one of its wealthiest ports and a center of culture.

In probing back to the origin of the Atlantis legend, the reality begins to look every bit as dramatic as the legend. While parts of the myth, such as an Africa-sized continent sunk in the Atlantic, are in conflict with the real-life geology of the ocean basins, other parts of the story, including the existence of a buried city with highly evolved architecture—even the Minoan custom of bull worship and the colors of rocks from which walls were made—match with spine-chilling fidelity Plato's description of a lost civilization that once held dominion over the eastern Mediterranean, and then vanished unexpectedly, and suddenly, at the height of its power. "There occurred violent earthquakes," wrote Plato in 347 B.C. "and a night of rain, and the island of Atlantis . . . disappeared, and was sunk beneath the sea."

When explorers first tunneled into the side of a Theran ravine and discovered rooms decorated with frescoes, when they located toilets and ceramic pipes and rattan beds, they assumed immediately that they'd been extraordinarily lucky and had stumbled upon the royal palace. Then, as streets and buildings—more and more of them—slowly came to light, they realized that this was how ordinary people lived. Baths in the homes of private citizens in so ancient a city? It was difficult to believe, but there it was. Whatever would the palace look like, if it was still intact?

Under the direction of archaeologist Christos Doumas, the excavations at Thera are revealing multistoried, exquisitely decorated buildings, complete with bathrooms and running water. Plumbing and other technologies emerging from the ash are of a complexity not seen again until the Islamic empire of the Middle Ages. The walls and streets are literally honeycombed with pipes.

The question that nags me is: since Thera is mostly desert today, and the record of fossil plants does not suggest that rain patterns in the Aegean were very different in 1650 B.C., where did all the water come from? It was fresh water—we know that much because there are no traces of salt in the sewers—and there was plenty of it. Was water tapped from springs on the vanished mountain? Did the Therans build aqueducts? Were they able to pipe water wherever they wanted it, into water towers and rooftop cisterns? The abundance of plumbing suggests that fresh water ran so plentifully on ancient Thera that it must have sluiced continually through the underground sewage system, flushing the city clean. Apparently there was water enough to waste. On this point the archaeological evidence is in agreement with Plato's account of the Atlanteans. His mythical island is said to have been blessed with hydrothermal vents:

And Poseidon, receiving for his lot the island of Atlantis, begat children by a mortal woman, and settled them in a part of the island, which I will proceed to describe. Towards the sea, half-way down the length of the whole island, there was a plain which is said to have been the fairest of all plains and very fertile. Near this plain . . . there was a mountain, not very high on any side. In this mountain there dwelt one of the earth-born primeval men of that country, whose name was Evenor . . . and he had an only daughter who was called Cleito. The maiden had already reached womanhood when her father and mother died; Poseidon fell in love with her. . . . He himself, as he was a god, found no difficulty in making special arrangements for the centre of the island, bringing up two springs of water from beneath the earth, one of warm water and the other of cold, and making every variety of food to spring up abundantly from the soil . . . They had fountains, one of cold and another of hot water, in gracious plenty flowing; and they were wonderfully adapted for use by reason of the pleasantness and excellence of their waters. They constructed buildings about them, and planted suitable trees; also they made cisterns, some open to the heaven, others roofed over, to be used in winter as warm baths; there were the king's baths, and the baths of private persons, which were kept apart; and there were separate baths for women, and others again for horses and cattle, and to each of them they gave as much adornment as was suitable. Of the water which ran off they carried some to the grove of Poseidon, where they were growing all manner of trees of wonderful height and beauty, owing to the excellence of the soil; the remainder was conveyed by aqueducts. (*Critias*, 113c–114a; 117a–c)

It begins to look as if the history of the world took an entirely different and unexpected path the day Thera erupted. Judging from their architecture, their frescoes and their intricate plumbing, the Therans were a graceful, refined and brilliant people. There is sadness here, sadness in the knowledge of a civilization that seemed on the verge of great deeds and never got the chance, simply because it happened to be located in the wrong place at the wrong time. It is possible that, had the island never exploded, the first expedition to the moon would have ascended from Crete, or Greece, or one of the Cyclades, before the birth of Christ, and that people might now be living in colonies at Alpha Centauri.

The island was, from every paleontological and archaeological indication, one of the most beautiful places on earth. The Egyptian pharaohs are said to have envied the Minoans their good fortune. That was their misfortune.

The island was also the most dangerous place on earth; but the Minoans did not know this. That was *their* misfortune.

Thera is now a ring-shaped string of islets whose central lagoon drops vertically, hundreds of feet, to life-giving hot water springs. The largest of the islets looks from the air like a curl of broken rib. Almost nothing grows on it, and there is a lost city under it, with canals and drainage systems, and apartments that look for all the world like present-day ocean-front condos. Plato, when he wrote about the sinking of Atlantis (*Timaeus*, 112), described "a land carried round in a circle and disappeared in the depths below. By comparison to what then was, there are remaining in small islets only the bones of the wasted body, as they may be called, all the richer and softer parts of the soil having fallen away, and the mere skeleton of the country being left."

The island, like Plato's Atlantis, is a broken skeleton of its former self. Its whole center is missing and flooded with sea water, and near its southern coast you will find one of the world's most hauntingly beautiful archaeological sites.

During the past week, I have clambered around on the thousand-foot-high rim of Thera and followed roadcuts through layers of ancient pumice with Christos Doumas, who, despite his fifty-five years and the onset of frequent allergy attacks (volcanic dust is beginning to bother him), can lead the way without losing his breath. Archaeologists work as hard as mountain climbers, and sometimes harder than ditchdiggers, he is fond of saying. Doumas has been digging for more than twenty years, uncovering Atlantis.

From atop the highest point on the rim, the circular shape of Thera Lagoon is immediately apparent. The circle is framed by andesite walls rising nearly a quarter mile out of the Aegean, the caldera of an explosion that hoisted two thirds of the island into the sky, destroyed the Minoan fleet and—almost certainly—created the legend of Atlantis. For reasons that will never be understood, the Greeks settled this place hundreds of years later. They named it Nea Kameni, the New Burnt Land. They did not survive the emergence of the black island in the center of the bay, or the quakes that accompanied it. A dozen civilizations followed. And every few centuries, a giant awoke to topple their buildings—all of them. Yet still the people came back. Now, a civilization in control of lightning and supersonic machines has erected new schools and stoplights next to Byzantine churches and mudwalled houses. It has sunk shafts into the very skin of Thera, and probed the lost city.

Here and there on top of the city, the volcanic ash and pumice are more than a hundred feet deep. In the uppermost layers of ash you will find Greek, Egyptian and Roman ruins.

The excavations began in 1967, and from the start it became clear to Doumas that he would never live to see his work finished. A single Theran room produced several hundred items, including bronze vases, tools, fossilized food and furnishings, all of which had to be cleaned, mended, photographed and catalogued. And, for all Doumas knew, there was a whole palace waiting to be discovered in this, the most completely preserved prehistoric site in the world. The city would have to be excavated by inches. Whole generations of archaeologists would live and die on Thera before the work was completed.

In twenty years, only one city block has been excavated. Test digs have shown that the streets run more than a half mile away, under churches and fig trees and the little farming community of Akrotiri, where I happen to be staying. At night, I actually sleep on top of the lost city. At the current rate of exploration, three hundred years of digging lie ahead. The excavation proceeds somewhat like open pit mining, except that it proceeds in slow motion, sometimes with paint brushes instead of shovels. Steel pillars, and a corrugated tin roof that runs level with the ground, have been erected over the entire excavation. The roof is necessary, for even the lightest rain might cause the buildings to crumble. But it does more than shut out the weather; it cuts off all the noise from the outside as well. I feel as if I am walking through a cave. I am reminded of a scene in that old movie *Journey to the Center of the Earth*. Near the end, James Mason and his companions enter a cave—roughly the same size as this excavation—and find the bakeshops and altars of Atlantis. What I remember most about that scene is Bernard Herrman's haunting score. And here, only a hundred feet from me, is what looks for all the world like the fictional bakeshop—storage jars of grain still in place—but there is no Bernard Herrman music, nothing except the scratching of my pencil and the riffling of pages. And when I stop writing— nothing. Absolutely nothing.

It is a mournfully beautiful silence. My footfalls echo back at me, and I know that I am walking through a world now so utterly dead that the sunlight under which the Therans played has long since crossed the Orion Nebula and the Pleiades, and is presently racing somewhere beyond the nearer rim of the galaxy.

Under the tin canopy, everything seems different. Time becomes elastic. In front of me is a structure—it looks like a multistoried apartment building—more than thirty-five hundred years old. It has bathtubs and toilets and plumbing and would not look out of place on a contemporary American street. Ten feet away is a support for the tin roof that covers the site, and beneath that support are traces of yet another lost city. They located it while digging a foundation for the pillar. Subsequent foundations have also penetrated through to the streets and walls of the earlier civilization—a buried city beneath a buried city. It is two thousand years older and it, too, seems to have been destroyed by the volcano.

Down here, time stretches to its outermost limits, and it seems an act of purest arrogance to believe that the Chrysler Building, or the Toronto Tower, or any other architectural works of our twentieth century will, after a millennium or two, be any less fleeting than a child's sand castle. Will even a square of sidewalk remain of Manhattan when it reaches Thera's age? Will the book I am about to write, and every other book of our time, cease to be even a memory? It squashes a man's ego. And strangest of all is to think that the very eruption that destroyed the city has protected it from rain and wind and looters—has preserved its story—while on the earth above, architectural marvels two thousand years younger have crumbled to lime in the open air.

Geography, oceanography and paleoclimatology tell us that in the autumn of 1628 B.C. the island of Thera was ripped open and blown apart and strewn about. Tidal waves spread outward from the rend in the earth, and archaeology tells us that the whole eastern Mediterranean was thrown into turmoil about this time. On Crete, the very heart of Minoan civilization faltered, stumbled and was soon taken over by mainland Greeks. On other shores, mass migrations seemed to commence everywhere at once. Smaller armed bands joined with larger armed bands, leaving burned villages and broken pottery behind them, taking what they could carry as they went. Egypt came through intact, but on the western shores of Turkey and in Syria, the Hittite Empire fell as wave after wave of invaders occupied Asia Minor. The Philistines, fleeing Crete, set up camp in Canaan (Israel) and were met by successive waves of Hebrew tribes fleeing the Nile (a mere footnote in world history, writ large by the Old Testament).

The Mediterranean world of the seventeenth century B.C. had little use for letters and writing. Outside of Egypt, no one seems to have written a detailed, eyewitness account of the upheaval, at least no one whose work survived very long. In Egypt, however, there were scribes and libraries, and priests to watch over them. A thousand years later, when Greek civilization was at its height, and its army had built outposts in Thebes and Sais, the scholars of Athens sailed down past Thera into the mouth of the Nile. And in time the Egyptian tale of a great cataclysm reached Plato, who transformed it in the fourth century B.C. into an epic—not as carefully researched history, but as a cautionary tale intended to demonstrate the vulnerability of civilization and to warn against the dangers of arrogance and corruption. It is easy to imagine that he embellished his story, either because the Egyptians had already embellished it for him or simply because he wanted to make it even more dramatic than it already was. Plato's Atlantis was not just an island civilization in the eastern Mediterranean Sea. The Mediterranean was too small, too close to home, too familiar. His audience might not have been impressed by anything so close at hand, and therefore less inclined to read about it. So the island culture swelled to continental giantism, and was moved somewhere mysterious and very far away, on the other side of Gibraltar, in the then uncharted and seemingly endless reaches of the Atlantic Ocean.

He must have known what he was doing, for a generation after his death the story had become one of the first great classics of literature. So widely was it reproduced and distributed that copies were simply bound to be preserved somewhere, and it became one of the few documents of the period that survives to this day.

But Aristotle dismissed it as a sort of fairy tale, suggesting that Plato had invented Atlantis to help him demonstrate a point, and then removed the continent once it had served its purpose. "The man who dreamed it up made it vanish," he said.*

Others did not dismiss the tale so easily. One of Aristotle's contemporaries, the Greek historian Cantor, went to Egypt around 300 B.C. to verify whether or not, as Plato had claimed at the beginning of his tale, the priests of Sais possessed ancient records of a great seafaring civilization that had vanished "in a single day and night of misfortune." The priests confirmed that such records still existed, though no one seemed to know the exact location of the lost world.

Whenever the economies of nations permitted it, expeditions went searching for traces of the sunken continent. When Spain

---

* According to Strabo, II and XIII, and Proclus in *Timaeum* 61a.

found the Aztec and Maya civilizations in eclipse during the winter of A.D. 1510, living in a land filled with crumbling pyramids, it was assumed that there, in Mexico, lay the remnants of the antediluvian world that had brought the art of pyramid building to ancient Egypt. But Mexico turned out not to be as ancient as the conquistadores had thought. By 1882 attention had focused elsewhere. In that year the British cabinet held a special session to discuss whether or not the Azores could be the mountain peaks of Plato's submerged land, and whether an Atlantis expedition vessel should be built and sent there.

While the Atlantis Plato described was technologically advanced by the standards of ancient Greece, it was quite primitive by today's standards. Yet ever since Plato, the legend has grown in surges, especially during the past two centuries, to the point that the Atlantis of twentieth-century mythology is technologically advanced even by today's standards and no longer bears any resemblance to the Atlantis Plato originally wrote about. Those who believe that the Atlanteans possessed electric generators, submarines and airplanes may be disappointed to learn that there are no submerged skyscrapers or airports on the Atlantic floor, and that what we are really talking about is a Mediterranean world with gold-inlaid bronze daggers, flush toilets and four-story buildings run through with sophisticated plumbing—but not the slightest hint of electrical wiring and light switches.

I first heard the story of Atlantis from my mother. It was the week of the Cuban missile crisis and at PS 23, in Flushing, we were having almost hourly air-raid drills, and classroom lectures about the atomic bomb. The teachers told us not to look at the Manhattan skyline on the way home from school because a great light might melt it and we would be blinded instantly. When I arrived home the grown-ups were talking about nuclear war. Someone said that sooner or later our whole civilization was going to blow itself up and disappear—just like Atlantis.

"What's Atlantis?" I asked.

My mother explained that a long, long time ago there had been a race of people called Atlanteans, who were "probably just like us." They even had plumbing in their homes, and they might have had airplanes; but their whole world sank and was lost in a single day and night.

Our neighbor added that it must have been a volcano or an earthquake, or perhaps even Noah's great flood. In both the Atlantis and Noah stories, he pointed out, the people destroyed were the people who had created civilization. Once they had been happy and sinless, but they became great and wicked and were destroyed for their sins. They were destroyed by water.

One of my teachers had said that if an atomic bomb exploded near Coney Island, it could empty part of the Atlantic Ocean and send a great wave crashing over Flushing.

"Did Atlantis have the atomic bomb?" I asked.

Today, the Atlanteans are said to have been too smart to have dabbled in nuclear power. Popular talk in this decade associates them with crystal and pyramid power instead. Despite the fact that oceanographers have found absolutely no evidence that a large land mass ever fell suddenly through the floor of the Atlantic, New Age enthusiasts continue to place it there, linking the Atlanteans to the pyramids of Egypt and Mexico (despite the fact that the pyramids of Mexico were built thousands of years after people stopped building pyramids in Egypt), to pyramids on Mars (despite the fact that Martian pyramids appear never to have been built at all), to UFOs and the Bermuda Triangle. There are presently dozens of people who claim to have been reincarnated from previous existences on Atlantis, which they date variously from 2000 B.C., 30,000 B.C. and 1,000,000 B.C., and place geographically everywhere from the Mediterranean to the Pacific. These are the sorts of things that inevitably come to mind when one mentions Atlantis—which is why serious archaeologists gathering round Thera wince at the mere mention of the name.

All of the modern folklore about ancient flying machines and pyramid power sounds fantastical, but the story at the bottom of the folklore is even more amazing because it is true. Now at last the tale of the lost sea people is being captured by archaeologists burrowing down into Thera and Crete. The deeper they dig, the more closely ancient Minoan lifestyles and religious customs (as revealed in wall-sized murals), the tall buildings equipped with running water, and the natural disaster recorded in their ruined walls, begin to dovetail with the customs, buildings and events described by Plato.

Three centuries of excavation lie ahead, difficult to imagine. I wonder what our own civilization will be then. If our culture survives long enough to complete the exploration of Thera, then all of the buildings that now surround me, all the beautiful frescoes emerging from every home—these things are but the beginning of a fantastic adventure; and I am grateful to have been here at the beginning, when in every direction lie mysteries, when there is no guessing what surprises lie beneath the next inch of ash.

I've heard it said that Thera is fast becoming one of the world's most fantasy-laden subjects, partly because archaeologists love to look over their shoulders at the deep time scales in which paleontologists and vulcanologists play, and we who stroll through the cellars of geologic time love to speculate about archaeology. For archaeologists and geologists, working together to reconstruct the story of a buried city, Thera is a joyous gathering place, a chance to play with the kids next door. The two disciplines have now begun to synthesize pictures of a Golden Age; a city of hanging gardens and tall buildings with pillars, a center of trade for more than a dozen nations. But the joy of discovery is tempered with the realization that Minoan Thera was merely an episode awaiting burial, first by waves of seething ash, and then by the weight of time itself. We see it as a beautiful dream, a remote Atlantis, no more.

I can pick up pieces of fossilized grain and loaves, then walk down Telchines Road and let my imagination follow my eyes across time. I've entered a bakeshop in a noisy, thriving Theran port, and then, only a moment later, smelled that shop—dry, ash-laden air, stagnant under the tin canopy of an archaeological preserve.

We never really know if we are piecing together the right story. But it's always fun to look at a fresco of a young lady, to hold a fragment of brick or a sliver of pottery in your hand, and to try to form from them an image of life when the House of Sorrows was freshly painted and still in the sun, when hanging gardens were everywhere, and when pipes conveyed streams of both hot and cold water. Irresistibly, you find yourself wondering about the people. You start fleshing out the lost souls of Thera, breathing imaginary life into them, and you understand a solemnity that is impossible to describe. If you were somehow able to look in on the last dance aboard the *Titanic*, you might get some feeling for what I am talking about. I am, in my mind's eye, living with ghosts at Thera. Standing on a street in a dead city, I imagine children coming out to play. It is a very haunting thing. I know what is going to happen. They don't.

There is, one knows not what sweet mystery
about this sea, whose gently awful stirrings
seem to speak of some hidden soul beneath.

—Herman Melville, *Moby Dick*

ATLAS
ADRIFT

Plato could not have known, when he penned his history of Atlantis in 347 B.C., that he was generating enough questions to keep historians, archaeologists, geologists, poets, philosophers and theologians busy for a very long time. More than two thousand years have passed, and we look back with puzzlement to an event that is, according to Plato, supposed to have occurred at least a thousand years before he was born. Was the story of the lost civilization really just a cautionary tale about men who dared to believe that they could tame nature, who dared to boast that they might even bring Zeus under their influence, and about how such unrighteous ambition proved to be their ultimate ruin? Was it all a mere fiction? Or was the Atlantis story, much like a game of Telephone played across fourteen generations of tellings and retellings, a distorted memory of something that actually occurred?

Plato, a friend and former student of Socrates, told his story in the form of dialogues between prominent Athenians of his day, including Timaeus and Critias, Plato's cousin. It is Critias who introduces the tale of lost Atlantis as it was related to him by his grandfather, also named Critias. Critias the younger was born in 460 B.C. and died in 403 B.C., when Plato was twenty-four years old. Critias the elder lived to the age of ninety and, according to Plato, heard the story from his father, Dropides, who in turn had heard it from Solon, the Athenian statesman, who, during a visit to Egypt in about 590 B.C., had met the old priest of the goddess Neith. The priest had access to the "House of Books," and he had told Solon the story of Atlantis.

According to Greek history, Solon did indeed visit Egypt near 590 B.C. Having just initiated a series of political and economic reforms in Athens, he decided to disappear for a while and let his new ideas take root in his absence. So he went exploring.

The incorporation of Greek architectural styles into Egyptian temples, combined with an abundance of Greek artifacts at archaeological sites along the Nile, suggests that Athenians were tolerated, if not welcome, in Egypt at the time of Solon's visit. Sais, the place Plato says Solon went to, was only ten miles from Naucratis, a major Greek outpost established on the Canopic branch of the Nile. Sais was the administrative capital of Egypt. Egyptian scholars were, at that time, looking back with keen interest on the long history of their forebears' forebears. The Great Pyramid at Giza was already two thousand years old and displayed scars of weathering and vandalism. The Greeks, by comparison, were a civilized people still in their infancy.

A man like Solon, whose overwhelming curiosity had sent him on what was, by today's standards, a long and difficult journey across the Mediterranean, would naturally have sought out the Egyptian priests, antiquarians and archivists. Communication was not a barrier to him, because the pharaoh had established a school of interpreters specifically for Greek visitors. In Sais, Plato claims, an old priest told Solon that Egyptian history streamed back through time so remote as to be incomprehensible to Greek scholars, who had little or no perspective on the past, except through vague myths about Minotaurs, or Phaëthon, or the strong and daring Titans, whom Zeus smote with thunderbolts and banished to a sunless abyss. "You have no antiquity of history, and no history of antiquity," the priest of Neith told Solon.

There have been, and there will be again, many destructions of mankind arising out of many causes. . . . There is a story which even you have preserved, that once upon a time Phaëthon, the son of Helios, having yoked the steeds in his father's chariot, because he was not able to drive them in the path of his father, burnt up all that was upon the earth, and was himself destroyed by a thunderbolt. Now, this has the form of a myth, but really signifies . . . a great conflagration of things upon the earth recurring at long intervals of time; and from these calamities the Nile, who is our never-failing savior, saves and delivers us . . . for which reason the things preserved here [in Egypt] are the most ancient. The fact is . . . that mankind exists, sometimes in greater, sometimes in lesser numbers. And whatever happened either in your country or in ours, or in any other region of which we are informed—if there were any actions noble or great or in any other way remarkable, they have all been written down by us [Egyptians] of old, and are preserved in our temples. Whereas just when you and other nations are beginning to be provided with letters and the other requisites of civilized life, after the usual interval, the stream from heaven, like a pestilence, comes pouring down, and leaves only those of you who are destitute of letters and education; and thus you have to begin all over again as children, and know nothing of what happened in ancient times, either among us or among yourselves. As for those genealogies of yours which you have just now recounted to us, Solon, they are no better than the tales of children; for, in the first place, you remember a single deluge only, whereas there have been many previous ones; and, in the next place, you do not know that there formerly dwelt in your land the fairest and noblest race of men which ever lived, and that you and your whole city [Athens] are descended from a seed or remnant of them which survived.

This power came forth out of the Atlantic Ocean, for in those days the Atlantic was navigable; and there was an island situated in front of the straits which are by you called the Pillars of Hercules [Gibraltar]: the island was larger than Libya and Asia put together, and was the way to other islands. . . . Now, in the island of Atlantis there was a great and wonderful empire, which had rule over the whole island and several others. . . . They had subjugated the parts of Libya within the Pillars of Hercules as far as Egypt, and Europe as far as Tyrrhenia [Italy]. (*Timaeus*, 22c–23c; 24c–d)

Plato tells us that Solon returned to Athens (Greek history tells us the same thing), and claims that he then told the story of Atlantis to Plato's ancestor Dropides.

Like the cautionary myth of Theuth and Thamus in Plato's *Phaedrus*, it is a fantastic story. But Atlantis has the curious distinction of bearing Plato's repeated assurance that it is "a fact and not a fiction."

And while Plato tells us that the story has been passed mostly by word of mouth along five generations of his forebears, he also makes a tantalizing reference (*Critias*, 113a–b) to a manuscript of Solon: "Before proceeding any further," says Critias the younger, "I ought to warn you that you must not be surprised if you should perhaps hear Hellenic names given to foreigners. I will tell you the reason of this: Solon, who was intending to use the tale for his poem, inquired into the meaning of names, and found that the early Egyptians in writing them down had translated them into their own language, and he recovered the meaning of the several names and when copying them out again translated them into our language. My grandfather had the original writing, which is still in my possession, and was carefully studied by me when I was a child."

More than two thousand years later, all we have is Plato's word that Solon's manuscript actually existed. The manuscript itself, if it ever was, did not survive. Did Plato ever see it? Or were Solon's notes of no more substance than a phantom continent in the mid-Atlantic?

SUMMER, A.D. 1968

Plato placed the Atlantean cradle on a lost continent west of Gibraltar, outside the mouth of the Mediterranean, and he sank it within the last few thousand years, in a single day and night. To put Plato to the test, given recent advances in deep-sea exploration, it should surely be possible to find traces of a continent, or even a Thera-sized island, falling through the floor of the ocean. There would be dikes and fissures on the bottom, thousands of cracks radiating hundreds of miles in every direction from wherever the island had been. They should still be glowing from the violence of their formation.

No light penetrates to the abyssal plains a mile and a half down in the open sea. And if a submarine's Plexiglas viewports should suddenly give way, the ocean would pour in with such terrific force that, in the space of a lightning stroke, your body is reduced mostly to individual cells. If you are the careless sort, there is always the consolation of knowing that nobody has ever died slow on the deserts of the deep.

On July 20, 1968, as the Soviets and the Americans, racing each other to the moon, began to recover from the *Apollo 1* fire and the *Soyuz 1* crash, the United States Navy was exploring a chain of dead volcanoes some fifteen hundred miles west of Gibraltar. Each of the mountains stood more than two miles high, though none of their peaks came within a half mile of the ocean surface. Plodding up the slope of a seamount, Woods Hole's inner spaceship *Alvin* looked like a giant sea creature with three glowing eyes. Inside its steel shell, Larry Schumaker and Robert Ballard looked out upon what appeared to be an alpine meadow covered with fresh-fallen snow. It was far from being fresh fallen, however. The "snow" was in fact the excreta and skeletal debris of plants and animals living in the sunlit waters above; and it had been accumulating on the bottom for millions of years.

A stiff half-knot "wind" was blowing from the south, lifting up a paper-thin layer of sediment and flinging it past the sub, as if to create an undersea blizzard. *Alvin* continued uphill, riding along the bottom on a single runner that prevented her hull from plowing into the slope and stirring up still more debris. She left a trail, like that of a one-legged skier gliding over virgin snow. The sediment was presently accumulating at less than an eighth of an inch every few centuries. A thousand years hence, no one in the United States Navy would remember why *Alvin* had been sent there in the first place. In all likelihood, no one would even remember the United States Navy; but *Alvin*'s trail would still be there.

As they neared the top of the volcano, Larry and Bob rode over a dune field. In places the snows had parted to reveal rubble coated with manganese. Larry pumped the propellers, made a few quick adjustments, then stopped for a look around. Some of the rubble resembled pieces of brain coral. And there were branching shapes just within range of *Alvin*'s floodlamps. They looked like deer antlers growing out of the bottom.

"Coral?" Larry guessed. "But that's impossible. It needs sunlight in order to grow. And we're almost a mile under. The sun never shines down here."

"I think I understand," Bob said. "A whole reef lived here, but it used to be up in the sun. That must have been a very long time ago. My guess is that a short way up this slope, if we dug down a few feet, we'd find the stumps of fossil trees—trees that once overlooked a beach."

"It sank, then. The whole island."

Indeed it had—but not in a single day and night. It could not possibly have been Plato's Atlantis. The floors of all the oceans in all the world are dotted with dead volcanoes, thousands of them, and all of them are sinking even as you read. It is a slow, stately sinking. Once a volcano stops squeezing up new rock through the earth's skin, once it ceases to create new land, it subsides naturally under its own weight. The lost island upon which *Alvin* parked during that summer afternoon of 1968 had, long before the Pleistocene Ice Ages, looked exactly like Hawaii: a huge seamount breaking the ocean surface. It need only have sunk inches per century to reach its present depth. And even Hawaii, a few million years hence, will crumble in silence in the absolute black of the deep sea.

Five million years lay under Robert Ballard's feet. As he peered through a snowstorm of particles swirling about dead antler coral, it occurred to him that there were many myths about an island civilization lost in the depths; but he knew that he would find no abandoned buildings here. This particular tract of land had slid beneath the waves long before Plato or Solon was born, long before the emergence of civilization, probably long before the emergence of man. If Bob wanted to see ruins on the ocean floor, he would have to wait about three million years, until the inch-per-century descent of Honolulu was complete.

No, there were no lost civilizations to be found on the bed of the Atlantic. It was all sinking volcanoes, nothing more. Just the same, they named the place Atlantis Seamount.

During that very same summer of 1968, the research vessel *Glomar Challenger*, equipped with powerful new drilling equipment, began seeking out the oldest ocean rocks. These would eventually be discovered at the foothills of continental shelves, near great folds that looked as if they were rolling under the shelves and passing beneath New York and London. None of these rocks dated back past the Age of Dinosaurs, which was puzzling, because the continents above were strewn with older rocks. Fossil bacterial reefs from Canada's Thunder Bay were chock-full of decay products from certain argon isotopes.* Radioactive argon breaks down so slowly that in order to account for such an abundance of decay products, the Thunder Bay rocks had to be almost

---

* The term isotope defines any of two or more forms of the same element, in this case argon, having the same number of protons in the nucleus, or the same atomic number, but different numbers of neutrons in the nucleus, or different atomic weights.

three billion years old. And the Allende meteoroid, still a year away from its fiery arrival in Mexico, would soon become the oldest piece of the solar system ever discovered, dating back 4.6 billion years. The puzzle lay in the fact that if you compressed the entire history of the world into four months, letting the earth and the Allende meteoroid form, say, on September 9, then the first dinosaurs arrived on Christmas Eve and the oldest ocean rocks did not appear until two days after the dinosaurs. Meteoroids and continents had somehow managed to preserve a record thirty times as old as the oldest ocean rocks.*

The youngest rocks of the oceans were found in such places as the mid-Atlantic Ridge, which runs like a seam through Iceland and reaches almost to the South Pole. From the ridge, rocks dating back only a few centuries were hauled onto deck. The mid-Atlantic mountain system lies at an average depth of one mile, with the oceans deepening east and west of it to three-mile basins. Here and there, mountain peaks actually break the ocean surface, as in the Azores—which had, in 1882, been put forth by Ignatius Donnelly, the "father of Atlantology," as the mountain tops of submerged Atlantis.†

Not likely. The mountains of the ridge have, for millions of years, been rising from the sea floor, not sinking into it. Moreover, the depth of "snow" or sediment covering the mountains—an hourglass or time probe, in effect—increases from a few inches to thousands of feet as one moves away, east or west, from the ridge. Some of the mountains are capped with dead coral reefs, and the depth at which a reef resides is directly proportional to its distance from the ridge.

A picture began to emerge from the *Alvin* and *Glomar Chal-*

---

* Humans then appear at 10:30 P.M. on New Year's Eve, and all of recorded history occupies the last ten seconds of December 31.

†Donnelly, a congressman from Minnesota, also claimed that Francis Bacon wrote all of Shakespeare's plays. In his *Atlantis: The Antedeluvian World* of 1882, the congressman suggested that the lost continent had spawned a "superior Aryan race" in Europe—an idea that, some fifty years later, would be eagerly snatched up by Nazi propagandists.

lenger expeditions. New ocean floor was being pushed up at the center of the mountain range. It then spread away in opposite directions, carrying dead volcanoes across thousands of miles of sea floor like parcels on conveyor belts. The farther they were from the spreading center, the older they were and the deeper they had subsided. Science was learning that the earth is not merely a ball of changeless, solid rock. Even on a scale of hours, it does not behave like an ordinary solid. As the oceans rise under the pull of the moon, the bedrock beneath our feet creaks and groans and rises with them, as much as a foot, every twelve hours.

We are forced to redefine the notion that rocks are changeless and immovable. Rocks, it seems, can stretch out and bend like warm plastic. Two hundred million years ago, Europe and North America began pulling apart to open the Atlantic Ocean. If you want to see right now how the Atlantic began, go to northeast Africa or Saudi Arabia and stand on the shores of the Red Sea. The planet is splitting open there. North Africa itself, carrying the Atlas Mountains with it, has plowed the island of Italy into southern Europe—like an out-of-control truck plowing a car into a snowbank—and pushed up the Alps. Climbers in Switzerland, Austria and Yugoslavia, as they ascend more than a mile above sea level, pass decks of ice-crusted rock made from the skeletons of sea creatures that once lived in the Mediterranean.

But even the Alps took millions of years to build, and their creation, as measured by the standard of human lifetimes, was a slow, gentle affair. Each million years they rose only hundreds of feet. Each thousand years they gained only inches.

If Atlantis was somewhere west of Gibraltar, more than twenty years of deep-sea exploration would have revealed signs of its sudden end. But west of Gibraltar you will find only gently rising and falling seamounts, and mile after mile of deep ocean snow, undisturbed since the day *Mososaurus* patrolled the waters. If there was any truth behind Plato's story, a continent sunk in the Atlantic was not part of it.

So Atlantis, if it existed, was located in a different place. The core of the story, as described to Solon in Egypt, as described to Dropides, as described to Critias the elder, as described to Critias the younger, as described to Plato, as described to us, appears actually to have occurred in the Aegean. The Minoans had developed an advanced civilization that, like Atlantis, became the strongest naval power the world had ever known, sovereign not only of its own island of Crete, but also of Thera, Naxos, Kea and all of the Cyclades. Crete, the seat of Minoan power, was the way to other islands. Its authority within the Pillars of Hercules extended over Greece, parts of Libya, Egypt and Europe as far as Tyrrhenia. This mighty empire did indeed have all appearances of coming forth out of the sea, and did indeed establish dominion over a region larger than Libya and Asia (insomuch as the western world knew of Asia at that time—which was not much, for the old mapmakers believed it to be smaller than Libya).

The abrupt decline of the Minoans follows close on the heels of the Theran city's burial. Unlike the Egyptian, Greek and Roman civilizations, the Minoans collapsed so quickly and so utterly— within fifty years of Thera's destruction—that history forgot them until ancient toilets, bathtubs and palace walls began turning up on Crete early in the twentieth century. Their technology survived in the Greek civilization that succeeded them, but they themselves remained little more than occasional recollections shadowed in Greek mythology: King Minos ruled the seas from Crete . . . Theseus entered the Labyrinth and slew the half-man, half-bull Minotaur . . . Daedalus, the Athenian architect who built the Labyrinth, made wings for himself and his son Icarus, to escape from Crete. The Daedalus myth seems to describe something akin to a hang glider. Daedalus conceded that Minos ruled the seas, but the king could not rule the air. So, on man-made wings the architect flew to freedom. Icarus is said to have ascended too close to the sun, melting the beeswax that bound his wings together. He plunged into the sea between Crete and Thera.

Between 1902 and 1909, the rediscovery on Crete of four cities and what had once been splendid palaces, with Greek ruins buried in the soil above them, revealed that a glorious and hitherto unsuspected civilization had existed before the age of Greek "Enlightenment." The picture that began to form suggested that the Greeks had in fact inherited many of their art forms, their metallurgy even their styles of pottery from the earlier civilization— which archaeologists quickly named Minoan.

In 1909, K. T. Frost, a professor at Queen's University in Belfast, proposed that the forgotten civilization was none other than the lost Atlantis described by Plato. Even without knowledge that the nearby island of Thera had buried at least one city and spread a death cloud over the Aegean and Mediterranean seas, there was, emerging from the soil of Crete, good reason for taking a closer look at the *Timaeus* and the *Critias*.

"The recent excavations on Crete have made it necessary to reconsider the whole scheme of Mediterranean history before the [Greek] classical period," wrote Frost in the *Journal of Hellenic Studies.*

> Although many questions are still undecided, it has been established beyond any doubt that, during the rule of the 18th Dynasty in Egypt, when Thebes [Karnak] was at the height of its glory [about 1650 B.C.], Crete was the centre of a great empire. The whole of seaborne trade between Europe, Asia and Africa must have been in Cretan hands, and the legends of Theseus [and Daedalus] seem to show that the Minoans dominated the Greek islands and the coasts of Africa.
>
> The Minoan civilization was essentially Mediterranean, and is most sharply distinguished from any that arose in Egypt or the East. In some respects also it is strikingly modern. The many-storied palaces, some of the pottery, even the dresses of the ladies [as depicted on Cretan frescoes] seem to belong to the modern rather than the ancient world.
>
> When Minoan power was at its greatest, its rulers must have seemed to the other nations to be mighty indeed, and their prestige must have been increased by the mystery of the lands over which they ruled, which seemed to the Syrians and the Egyptians to be the far West, and by their mastery over the water—that element which

the ancient world always held in awe. Strange stories, too, must have floated round the Levant of vast bewildering palaces, of the [recently excavated] bull arena, of sports and dances, and above all of the bull-fight. The Minoan realm, therefore, was a vast and ancient power united by the same sea which divided it from other nations, so that it seemed to be a separate land with a genius of its own.

By 1909, archaeological evidence had already shown signs of widescale destruction in Knossos and its neighboring Cretan cities. Wood, once it is carbonized by fire, does not decay in the soil; and most of the excavated rooms seemed to be filled with burned wood, overlaid by rubble from upper stories, which was in turn overlaid by a mixture of grit (not yet identified as volcanic ash), the accumulated debris of nearly four millennia of sprouting, shedding and decaying trees, stream-washed sediments and the residuum of a dozen civilizations that happened to be passing through. Frost, regarding a natural catastrophe capable of sinking an island civilization as too fantastic to be true, dismissed that part of the *Critias* as a fiction invented by Plato to embellish the story. He attributed the sudden eclipse of Minoan power, and its subsequent replacement by Greek power, to a raid on Crete: "As a political and commercial force, therefore, Knossos and its allied cities were swept away just when they seemed strongest and safest. It was as if the whole kingdom had sunk into the sea, as if the tale of Atlantis were true."

Frost had already seen evidence of elaborate bathrooms, and the great Cretan harbor, where shipping and merchants must have come from all ports. The wreckage of Minoan Crete echoed scenes in Plato's *Critias*:

The docks were full of triremes and naval stores, and all things were quite ready for use. . . . The entire area was densely crowded with habitations; and the largest of the harbors were full of vessels and

merchants coming from all parts, who, from their numbers, kept up
a multitudinous sound of human voices and din and clatter of all sorts
night and day. (*Critias*, 117d–e)

"The bathrooms, the sea people, and the solemn sacrifice of the
bull: these things, as described by Plato, were thoroughly Minoan," said Frost. "But when we read in the *Critias* how the bulls
of Atlantis were hunted in the temple of Poseidon without weapons but with staves and nooses, I have an unmistakable description
of the bull ring at Knossos, the very thing which struck foreigners
most and gave rise to the legend of the Minotaur."

There were bulls who had the range of the temple of Poseidon; and
the ten kings, being left alone in the temple, after they had offered
prayers to the god that they might capture the victim which was
acceptable to him, hunted the bulls without weapons, but with staves
and nooses; and the bull which they caught they led up to the pillar.
(*Critias*, 119c)

As Frost considered the similarities between Plato's Atlantis
and Knossos, Minoan tombs were being found intact, unmolested
by looters. One tomb yielded teacup-shaped vessels with scenes
of bullfights engraved on their sides. The bulls were shown snared
in ropes. Similar scenes were depicted on the palace walls of
Knossos—scenes that differed from all other bullfights known
through history in precisely the point Plato had emphasized: no
weapons except ropes were used.

The Minoan tomb vessels agreed with Plato's story on one other
point. According to Plato, the bull, once captured in the temple
of Poseidon, was led to a sacred pillar in the middle of the island,
on which the laws of Atlantis had been inscribed, and there the
animal was sacrificed:

They cut its throat over the top [of the pillar] so that the blood fell upon the sacred inscription. When, therefore, after slaying the bull in the accustomed manner, they proceeded to burn its limbs, they filled a bowl of wine and cast in a clot of blood for each of them; the rest of the victim they put in the fire, after having made a purification of the column all round. Then they drew from the bowl in golden cups, and pouring a libation on the fire, they swore that they would judge according to the laws on the pillar. (*Critias*, 120a)

The Atlanteans, according to Egyptian legend (as passed down by Plato), hunted a sacred bull with ropes, then drank its blood from golden cups. The Minoan tomb cups not only depict bulls being hunted with ropes; they are also made of solid gold. If the bull ceremony, as described by Plato, is accurate, then the Athens National Museum may now have on display the very same cups from which the blood was consumed.

Plato's Atlantis story seems to describe customs and events now being mirrored in archaeology and geology. If some essentials of the Atlantis legend are true, how then did the lost civilization come to be located on the wrong side of the Pillars of Hercules?

In 1939, Spyridon Marinatos, Christos Doumas's teacher and the first director of the excavations here at Thera, put forth a theory in the archaeological journal *Antiquity*. He attributed the Minoan decline to the volcanic eruption of Thera (about 1600 B.C.). "This at once removed the weak link in Frost's 1909 argument," wrote Doumas in 1983, "because Crete's collapse could supposedly be shown to have been sudden. Marinatos himself later admitted to a core of truth in Plato's legend. According to him, the Egyptian priests who spoke to Solon about the end of Atlantis were referring indirectly to the late Bronze Age eruption of Thera: the complete destruction of the Minoan fleet* . . .

* As many theorists have put forth, by a death cloud hot enough to set wood and clothing and skin ablaze, or by tidal waves that smashed ships in their harbors.

resulted in a sudden break in Minoan-Egyptian contacts. Word
of some of the effects of the eruption—complete darkness, ashfall,
poisonous gases, tsunamis—must have reached Egypt." J. V.
Luce, professor of classics at Dublin's Trinity College, believed
that what the eruption did to Thera was unimportant compared
to what it did to the Minoan archipelago surrounding the island.
("A brilliant and refined culture foundered under the brutal impact
of Theran volcanism.") Moreover, Doumas continued, "the rumor that an island had sunk into the sea would certainly have
reached Egypt very quickly. Thus, according to Marinatos, it was
not difficult for the Egyptians to confuse Thera with Crete, imag-
ining that the submerged island was none other than the large
island civilization (Crete) with which they had suddenly lost
contact."

Marinatos often wondered what records the priests of Lower
Egypt must have made available to Solon. Even today there are
papyrus documents surviving that make reference to early contacts
between Egypt and a place called Keftiu, which is generally accepted to be Minoan Crete. Keftiu is first mentioned in Egyptian
writings dating back to the third millennium B.C.:

> Keftiu and Isy [Cyprus] are under the awe of thee. I cause them
> to see their majesty as a young bull. Firm of heart, sharp of horns,
> who cannot be felled.

It fades from reliable records before the third quarter of the
sixteenth century B.C.

"If Solon had inquired more particularly about Keftiu," writes
Professor Luce in *Lost Atlantis*, "he would have been told that it
was an island far away in the west. The Ipuwer papyrus uses the
phrase, 'as far away as Keftiu.' . . . If Solon pressed them on the
ultimate fate of Keftiu, the priests could certainly have told him
that it disappeared from their records around the middle of the
XVIII Dynasty [around the time of Tuthmosis III, Hatshepsut

and Amenophis II.]* "As for the location of Atlantis, assuming Keftiu was the source of the legend, Marinatos liked to point out that the classical Greeks, having expanded into and colonized the Mediterranean (presumably filling a vacuum left by the Minoans), began to displace legendary events across the Pillars of Hercules, into the equally mysterious and seemingly infinite Atlantic Ocean. So Solon might have placed Atlantis there even if the Egyptians did tell him otherwise. The island was said to have been larger than Libya and Asia (interestingly, Crete's area of influence seems to have been as large), and Solon would have been troubled by the implausibility of accommodating such an enormous island (rather than an island's range of influence) within the Mediterranean.

But if we are to place Thera and Crete at the center of the Atlantis legend, we must also explain the date of submergence given by Plato: nine thousand years before Solon. Again, there is a historic tendency for ages to be exaggerated beyond credibility the further back we look into legendary time. In the Old Testament, for example, as we leaf backward from the time of its writing (at least two hundred years after Moses, near 1400 B.C.), we read that Moses died at the age of 120, Jacob at 147, Isaac at 180, Terah (Isaac's grandfather) at 205, Eber (Terah's great-great-great-grandfather) at 464, Shem (Eber's great-grandfather) at 600, Noah (Shem's father) at 950, and Methuselah (Noah's grandfather) at 969. When we look back past the early patriarchs of the ancient Jews, to the even more ancient Babylonian legends,

---

* Amenophis II, who ruled about 1600 B.C., is the last pharaoh whose tomb hieroglyphs make reference to Keftiu. On the walls are paintings of foreigners bearing Minoan objects. The Karnak tomb of Rekhmire, vizier to Tuthmosis III (who ruled during the fifty years preceding Amenophis II, and has long been regarded by biblical scholars as a pharaoh who oppressed Moses' people), also bears written references to Keftiu, beneath paintings of men carrying typically Minoan bowls and rhytons (fragments of which were actually found buried in the tomb). But their hairstyles do not resemble the locks and half-shaven heads depicted on many Minoan frescoes, and their kilts, which were originally painted to show Minoan styles, were painted over again to show a longer, more ornate style characteristic of early mainland Greece. The repainting seems to reflect Egyptian awareness of a shift of power in the western sea.

individual kings are said to have ruled for tens of thousands of years.

There have been attempts at attributing the nine-thousand-year dating (before Solon's visit to Egypt) of the Atlantis deluge to clerical error—as, for example, to suppose that Solon confused the Egyptian symbol for 100 with that for 1000. Nine hundred years before Solon would be a much more acceptable date for the end of Atlantis (about 1500 B.C.), in fact more or less the time of Thera's destruction; but such explanations are unconvincing and unnecessary. If a legend, passed down for nearly a thousand years before it reached Solon, and for an additional two hundred years before Plato finally committed it to paper, can, through successive retellings, expand an island culture into a continental one and dislocate it west a thousand miles, it is easy to believe that Plato's family, or the Egyptians, or both, stretched its age as well.

J. V. Luce points out that the original name for Crete—Keftiu— is probably derived from a root meaning "pillar," and that a western island containing a "sky pillar," that is, a lofty mountain helping to support the dome of the sky, would fit neatly into the frame of Egyptian Bronze Age mythology. A goddess was said to hold the heavens on her back, and the sky pillars were the four points at which her arms and legs touched earth.

"It is even possible," says Luce, "that the Egyptian priests had some recorded material about the worship of sacred pillars which was so prevalent in Minoan Crete." An alien culture worshiping sky pillars, he noted, would have been of great interest to Egyptian theologians—an affirmation of their own beliefs. "Imagine Solon's reaction when confronted with this sort of information about ancient Keftiu. He could not have failed to associate it with the myth of Atlas, 'the Titan' who held the sky on his shoulders."

According to Plato (*Critias*, 114b), Poseidon had five pairs of male children with Cleito, the mother of Atlantis: "And the eldest, who was the first king, he named Atlas, and after him the whole island and the ocean were called Atlantis."

"But the name Atlantis is a most deceptive guide," suggests

Luce. "In Greek, Atlantis and Atlantic are adjectival forms of Atlas, meaning '(the island) of Atlas' and 'the sea of Atlas' respectively. . . . So if you decide to use the name of Atlantis as a clue to its location, you must consider what was the original (Greek) location of the mythical Atlas. Now Atlas may once have been located well inside the Mediterranean before the gradual extension of Greek geographical knowledge pushed him to the west and located him on the High Atlas range in Morocco. . . . The name Atlantis Sea for a portion at least of what we call the North Atlantic was in use a generation before Plato was born. It first appears in Herodotus in the form "the so-called Atlantis Sea," and seems to have developed without reference to the Atlantis legend. Herodotus was aware . . . of a tribe called the Atlantes, who dwelt round an oasis in the [Moroccan] desert far to the west of Egypt. They derived their name from a mountain called Atlas, which they regarded as a 'pillar of the sky.' It was, as befits such a pillar, 'narrow and completely circular, and so high that its summit could not be seen.' "

Which brings us back to Crete's original name, and to the blurring effect that nearly a thousand years of prehistory had on places and events that, through Egyptian eyes, were far away in the west. Keftiu translates as either "the island of Keft" or "the people of Keft," depending on which determinitive is added to the Egyptian hieroglyph. The root *keft* has come into the twentieth century as the word "capital," and in the Old Testament Keftiu is Caphtor, the capital of a pillar, whose people (Minoan Cretans?) are the Philistines.* Bronze Age Egyptians might have regarded the mountains of distant Keftiu as the place where one of the four "pillars of heaven" touched earth. Luce suggests that Solon translated Keftiu as Atlantis—the island of Atlas—and, since the in-

---

* The identification of Keftiu with the Biblical Caphtor, Luce explains, is emphasized by the Sargon of Akkad tablet, which refers to Kap-ta-ra as lying "beyond the Upper Sea" (the Mediterranean). Biblical passages connecting Crete (Kaphtor) with the Philistines include: Genesis 10:14 ("and the Caphtorim whence came the Philistines"), Amos 9:7, Jeremiah 47:4.

habitants bore the same name, he called them the descendants of Atlas.

"This must have seemed to him a very reasonable equivalence," says Luce. "The name would have scanned well in the epic poem Solon planned, and 'the island of Atlas' would have the right sort of mysterious and western flavor that his plot required. . . . There is, I think, great plausibility in [Frost's 1909] contention that 'Solon really did hear a tale in Sais which filled him with wonder, and which was really the true but misunderstood Egyptian record of the Minoans.' "

SUMMER, A.D. 1932

On the Thera-facing, northern shore of Crete, at a place called Amnisos, Professor Marinatos climbed out of the pit, stood on its edge, and took a grandstand view of the last three weeks' excavations. A building had stood here, just a short walk from the beach. It must have been a wealthy man's summer villa, but only its foundations had survived. This was perfectly normal. Archaeologists were used to having nothing more than foundations to work with. This foundation, however, was unusually out of line. Stones more than three feet across had been kicked about, and in the bottom of the rectangular depression was a thick layer of sand, ash and pumice pebbles—feather light, frothy pebbles that actually floated upon the sea, and occasionally washed up on beaches of the Greek islands.

A picture began forming in his mind. The whole top of the building had been swept away in an instant, erased by a tidal wave that spared only the disrupted foundation. The grit and pumice were deposited later, over the course of weeks, or years, by rainwater washing into and draining out of cavities in the ruined structure. The water paused there, dropping its sediments.

The orthodox explanation for all of this—for the sudden destruction and abandonment of the villas and palaces of ninety

Minoan towns and cities—was that foreign invaders had descended upon Crete.

Marinatos just didn't see it that way. The Minoans were a naval power—the world's first and only navy. The sea was Crete's wall, and ships reinforced the barrier. To invade Crete, you would have first had to punch a hole through the Minoan Navy—if not somehow eliminate the fighting force altogether. To then overthrow ninety towns and cities, your own ships had to carry a landing party of hundreds, if not thousands, of soldiers.* Who besides the Minoans possessed such a navy? Marinatos wondered. There was no one else. No one at all.

No country could possibly have mounted an invasion against Crete unless the island's navy was already gone or severely disabled. Its citizens, too, would have had to be in a weakened condition, as from a plague, to yield to such an invasion. The downfall of the Minoans was not a tale of struggle and displacement, but one of simple replacement by people who moved in and took command after Crete had ceased to be a major power. It had to be so.

In the excavation next door, barely a quarter mile up the road, and more than fifty feet above sea level, Marinatos had seen walls that seemed to have collapsed in an unusual way. Rectilinear blocks more than six feet long had been moved about in a manner that suggested the sucking and dragging motion of water receding

---

* It has been argued that fewer than two hundred Spaniards were able to overthrow Mexico's Aztec Empire; but this was a historic anomaly. Cortez and his soldiers just happened to arrive in the year that a terrible war with the gods and destruction on earth had been foretold. The empire was then in a state of diminished morale comparable to America during the depression of the early 1930s. Moreover, the Spanish conquest of Mexico was one of high tech meets low tech, high tech rapes low tech. It was cannons and armor plate against obsidian daggers. It was also a biological conquest of smallpox, measles and chicken pox over people who had never been exposed to these illnesses. The invaders, protected by immunity acquired at the price of uncountable deaths in preceding generations, rode through unaffected while whole cities perished, rode through on horses—which no one in North America had seen before. (They *must* have seemed like gods.) In the Bronze Age Aegean, no nation had such advantages over the Minoans. It was the Minoans, in fact, who possessed superior technology.

in a huge mass. The whole upper part of a vertical block was missing. It appeared to have been snapped off. Recently, a digger had discovered the missing piece jutting out from under a bush. The two broken halves matched perfectly; but the upper part had been deposited more than three hundred feet away from its foundation.

Invading people don't do that, Marinatos judged. They may rape and loot and burn, they may even smash a few monuments; but they do not go to the trouble of breaking ordinary stone walls apart, and hauling half-ton pieces three hundred feet.

There was also a great deal of carbonized wood in the house— evidence of extensive fire damage, and this point seemed to contradict Marinatos's tidal-wave theory: wet houses don't burn. However, during recent weeks, he had begun to read everything he could find about the eruption of Krakatoa, west of Java, in 1883. By comparison to what he was learning from his inspection of the hole in Thera, Krakatoa had been a relatively mild eruption, burping a mere eight cubic miles of debris into the sky. When rescuers arrived in the community of Tyringen, thirty miles across the sea from Krakatoa, they discovered a town overturned by a wave. The water, as it marched through Tyringen, tore loose the lighter homes and carried them away in pieces. Glutted with debris, the wave had deposited drifts of rotting lumber and cloth and flesh two miles inland. A ship lay in the middle of the moraine, thirty feet above sea level. The rescuers also observed that most of the debris showed signs of having been burned. Marinatos theorized that the fires must have been started by lamps, which the inhabitants had lit to counteract the pall of darkness from the eruption. Not until the exquisitely documented Mount St. Helens eruption of May 18, 1980 (which blew a relatively insignificant half-cubic-mile hole in the earth), would anyone learn that volcanic dust can cut across the ground faster than a tidal wave. The very air that surrounded Mount St. Helens, made heavy with ash, flowed like water across the land, pushing down 3.2 million trees, all in the same direction, as if they were blades of grass washed over by a flash flood, except that this flood moved faster than a plane, and was hot enough to melt flesh and ignite wood ten miles away.

Marinatos could not know these things, as he stood in the beachfront ruins seventy miles from Thera, trying to form for himself the image of a black curtain going up in the north, reaching twenty miles high to the very edge of space, curdling the sky; and stirring out there, rising up on its haunches, he could see the tsunami. It got taller every second.

Was this the bull from the sea that was sent to plague Minos? Marinatos wondered. Have we here the grim historical reality behind Plato's words (*Timaeus*, 24d): "There occurred violent earthquakes and floods; and in a single day and night of misfortune the island of Atlantis disappeared in the depths of the sea"?

And the first angel sounded the trumpet, and
there followed hail and fire mingled with blood,
and it was cast upon the Earth; and the third
part of the Earth was burnt up, and the third
part of the trees was burnt up, and all green
grass was burnt up . . . and as it were a great
mountain was cast into the sea; and the third
part of the sea became blood, and there died
the third part of those creatures that have life
in the sea, and the third part of the ships was
destroyed.

—Revelation 8:7–9

T   h   r   e   e

A
KILLING
WIND

~~~~~~

AUTUMN, A.D. 1989

Volcanoes, like snakes and comets, have come down through history and legend with a bad reputation. Mention volcanoes to most people and they think immediately of Krakatoa, or Pompeii, or poor Harry Truman at Mount St. Helens. But if you've sailed, as I have, with Bob Ballard and Jean Francheteau into the perpetual darkness two miles below the Pacific, you'd see immediately that volcanoes are not merely bringers of tidal waves and fiery death. They are also bringers of life. Indeed, there is every indication, from laboratory experiments, that hot volcanic rocks were needed, about four billion years ago, to concentrate amino acids into little clumps and strings of proteinaceous spheres, into protocells—things that eventually became us. It is very likely that without volcanoes there would be no life on earth today.

I remember, about three years ago, floating over one of the volcanic seams in the earth, watching a rack of TV screens as Woods Hole's inner space probe, *Argo*, zeroed in on an active

53

vent. All of us scientist types had brought along 35mm cameras with fancy attachments, whereas the woman from *National Geographic* used a little pocket Kodak. Just a few yards away from the robot was a dense stand of red-tipped flowers, twelve feet high, that we had hastily dubbed "the rose garden." But they were not really flowers. They were giant worms that live by inhaling sulfides from within the earth. Volcanoes nourish them, and they live directly on the cracks in the world, evolving in strange ways. They lack mouths and anuses, or any traces of what we might call digestive systems. They are kept alive by bacteria that dwell in their tissues. The bacteria absorb volcanically vented sulfides and, by slowly burning or oxidizing them, help the giant worms to manufacture their own food in much the same way plants up here on the land use sunlight.

It is a fascinating thing, when you come right down to it: animals that look like and live like plants, in a black world where the entire food chain is based on sulfides. On the beds of all the oceans, wherever we find live volcanic springs, the sulfide-metabolizing bacteria thrive so richly that they literally carpet the bottom. Crabs and fish graze on them. Oases sprout seemingly out of nowhere, on the deserts of the deep. I've seen fields of clams—their shells more than a foot wide—and giant white crabs, and things we could not possibly identify. At one of the volcanic oases, we saw something that looked for all the world like a honeycomb, yet certainly there are no bees on the bottom of the Pacific. Submarine pilot Ralph Hollis reported seeing a sliver of metal about two feet long, shaped like a maple seed pod (the kind children stick to their noses to pretend they're Pinocchio). He thought it looked like a piece of scrap dropped by a passing ship, but when he went to grab it with a robot arm, it fluttered away like a bird. He flew the minisub over two more of these creatures, and each time they proved too fast for him, though I begged for a specimen. Hollis recalled chasing after something very much like them two and a half miles down on the floor of the Atlantic, near the wreck of the *Titanic*.* But he cannot be sure that they are even ancestrally related. And he really does not much care. On almost every dive, submarine pilots see creatures that no one has seen before. They've become used to it.

There is nothing in the world quite like eating microwave popcorn, looking out—through a robot's eyes—at a bouquet of giant tube worms and listening to the Beatles singing "Octopus's Garden." And the landscape: evil-appearing and blacker than a mineshaft beyond the probe's lights, cracked bulbs of volcanic glass immediately below the cameras, silted over with sulfides and bacteria, the rocks all black and gray, with occasional streams of sulfurous yellow. And just barely within range of our search lamps were chimneys jetting water heated far beyond boiling, almost hot enough to melt lead. If unleashed here in the ruins of Thera, or anywhere else near sea level, water that hot would normally explode into searing white vapor, but under the mountainous pressure of more than two miles of overlying ocean, it merely rises like sulfur-laden chimney smoke, and makes an oasis. Even the largest, most violent rends in the earth seem to have erupted down there only as gentle pools and rivers of lava. Explosions must be very rare, because the miles of water press down with a force equivalent to the thrust coming out the back of a space-shuttle engine. It is only up here, in the sunlight and the thin air, that volcanoes have the power to catapult steam and boulders, to blow enormous holes in the earth, and to kill on a grand scale.

Those gently smoking cathedrals are only a small part of a hydrothermal vent system that literally girdles the planet. The volcanic spreading centers run, much like the seam on a baseball,

---

* Apparently, creatures living near the vents are constantly setting eggs and larvae adrift. Most die without ever finding a food source. In 1912 an amazing thing happened. The *Titanic* fell out of the sky, bringing with it miles and miles of edible deckwood, and storerooms full of shelled walnuts and eggs and beef. The desert bloomed around her, became an undersea oasis that endures to this day. Similar oases are now being found around the skeletons of whales who have dropped two and a half miles to the bottom of the sea.

over forty thousand miles under the Indian, Pacific, Arctic and Atlantic oceans. Not many people appreciate that there are more volcanoes on the ocean floor than on the continents. Since humans have lived on this earth, fully a quarter of the water in the oceans has seeped down into the crust and been regurgitated through life-giving vents.

When you come to think about it, that's a lot of heat being stolen out of the earth by water. While the spreading centers and hot spots are active, continents are pushed apart (at about the rate fingernails grow) and undersea mountain ranges are built up. The ranges are so vast that they could easily accommodate the Rockies, the Alps and the Himalayas in a small corner. In effect, the ocean floor swells, pushing the oceans themselves up and spilling water onto the continents. About sixty-five million years ago, during the Age of Dinosaurs, most of North America was a major sea. Europe and Asia, and Africa too, were flooded. The larger ratio of water to land surface area produced a very mild climate, to which the dinosaurs must have become keenly adapted. Then the swelling began to subside. Probably, the engine that drives sea-floor spreading had given up too much of its heat to the oceans. The whole process slowed. The undersea mountain ranges began to sink under their own weight, and water drained back from the continents to fill the gap. Falling sea levels exposed larger continental boundaries, strangled major oceanic currents and trade winds, and caused a complex deterioration of climatic equability that might have been the start of, or at least a major cofactor in, the dinosaur extinction.*

* A close examination of the Ice Ages, especially the period between A.D. 1300 and 1800 known as the Little Ice Age, coincident with a recorded absence of sunspots, suggests that our sun has been behaving strangely over the last seventy million years or so. The Little Ice Age, which brought grief to George Washington at Valley Forge, routed the Vikings and fueled Europe's textile industry, turned the Baltic Sea into an unbroken ice field during winters of the fourteenth century, and allowed packs of wolves to cross from Norway to Denmark on a bridge of solid ice. A 115,000-year-old pollen sample from a French peat bog is particularly alarming. A forest of fir, spruce and oak gave way to Arctic pine barrens and tundra during an interval that, judging from radiometric dates above

Now there's an interesting paradox: too little volcanic activity may be even more devastating than too much. Oceanographer Rodney Batiza tells us that if we measure the ages and numbers of volcanoes per unit area on the Pacific Basin, we'll find that during the Age of Dinosaurs there was far more activity along the ocean's seams than there is today. The number decreases through the Great Dinosaur Vanishing Act, and then slowly increases during a period of rising sea levels (more spillage onto the continents?) and milder climate about fifty million years ago. Apparently, heat began to seep up again from the earth's mantle, causing more deep-sea mountain building, but the number of volcanoes per unit area decreases through a subsidence of sea levels ten million years later (amplified, perhaps, by the growth of polar ice caps), peaks again during a period of global warmth and continental flooding near fifteen million B.C., and then declines into the Pliocene Epoch, during which the Mediterranean Sea dried up and Thera emerged.

Today, in the center of Thera Lagoon, you will find a small island. It rose from the sea centuries ago. It stands about a thousand feet high and little more than a mile across. Here and there, hot water gurgles up through vents, dark brown and full of sulfides. It is a black, jumbled landscape of sharp-edged boulders and turbulent water, something out of the river Lethe in Dante's hell. In its center is a cone-shaped depression more than a hundred feet deep. Two days ago I descended into the bottom of the cone. I found a fumarole from which a steady stream of hot vapor was rising. The hole was only about six inches wide, and yellow sulfur crystals had condensed around its edges. They were clear-cut and

---

and below the pollen transition layer (which show no discernible time change) and from the extreme thinness of the layer itself (less than one eighth of an inch), was terrifyingly small: the forest became tundra in less than a century. In more ancient times, both sun and earth seem to have turned against the dinosaurs. The now-popular story goes something like this: after sea-floor spreading and solar output had eased off and the last of the dinosaurs were standing knee deep in snow and asking, "What else could possibly go wrong?" one of them looked up and saw an asteroid about to hit the earth.

needlelike—a delicate contrast to the broken andesite all around me. Gloria, my wife, took only two or three pictures of me (for scale) on the bottom—from about ninety feet above me, and at least five paces back from the edge of the cliff. (I could not coax her down to the edge for a clearer shot. I guess she has more sense than I do.) I had begun to chop off a cluster of sulfur crystals when something grumbled. It was a barely perceptible jolt, the voice of the earth's crust, a reminder of when and where I was. And there it was again! Time to leave, time to get back to the boat. Inside the throat of a volcano is no place to be testing the limits of one's luck. As I ascended, rocks began falling loose and tumbling down the sides of the cone. I had to sidestep a block of andesite. It bounced down and crashed.

Someday I'll learn to stop hanging around volcanoes; but not before I get down two miles to my favorite spot, the mid-Atlantic Ridge, in Woods Hole's minisub *Alvin*. I have to admit, however, that much as I love *Alvin*, I do not feel quite as invincible as I might have felt about three years ago. There was a time before *Challenger* when I was fool enough to believe that our machines were more powerful than nature itself, that there was something truly godlike in the universe—and it was us. Try hovering over a volcano in *Alvin*'s titanium shell; and pray that a stream of scalding water does not crack or melt the Plexiglas viewports. Then you will know—*really know*—that we can sometimes cheat nature, if we are wise, and pay attention; but we can never truly tame her. Our proudest marvels of technological achievement are overwhelmingly frail by the standards of the universe. Even *Alvin* is no challenge to a volcano. When the earth's crust speaks, we must obey and run.

SPRING, A.D. 1980

Harry Truman was thirty-one years old when he first arrived at Spirit Lake. That was in A.D. 1927, and during the half century that followed, he had managed to carve a living out of the wilderness. By 1980 he was operating a lodge in the very shadow of Mount St. Helens. He knew every hiking trail through what had become some of the Northwest's most popular forestlands. Though the onset of frequent tremors had rattled windows and shaken chips of concrete from the steps of his lodge, he refused every order to evacuate. He was a man trapped by his past. The last people who saw him alive remember him watering his lawn and repainting the lodge, in preparation for the next season's visitors.

Harry Truman never understood that he was living under sentence of death, or that an important piece of the Thera puzzle was about to fall into place. Mount St. Helens would soon become the most photographed, best-documented volcanic explosion the world had ever known. Scores of automatic cameras, remotely operated gas sensors, tiltmeters, seismic listening devices and temperature gauges had been planted within a twenty-mile radius of ground zero, where, on the morning of May 18, 1980, a half cubic mile of mountain—slabs of rock, ascending magma, pressurized ground water and blocks of ice—flashed to vapor and dust and began to swell over the sky. Four hundred million tons of rock became a superheated powder that spread throughout the surrounding atmosphere. The resulting death cloud was mostly air made heavy with ground-seeking bits of pulverized rock and, as such, it behaved more like water than like a gas. But this "water," as it tried to settle back to earth, was being hurled away from an explosion center far more violent than a nest of hydrogen bombs. The cloud spread huge and globular: in fifteen seconds it was three miles across. Three seconds later its diameter was four miles, and its northern edge burst down toward Truman's lodge, only three miles away.

Harry Truman probably felt nothing more than a jolt running through the ground. He neither saw nor heard, and certainly did not have time to comprehend, what happened next. The fist of rock-laden air went through roof and walls and floorboards as if they did not exist. Closing around his pink Cadillac, his 1883 player piano and his sixteen cats, it flung them out over Spirit Lake, horizontally in the terrestrial din. Within the cloud, dull orange light emanated from every direction. Flecks of dust glowed red and white hot, like sparks in a furnace. Harry Truman's brain almost had time enough to register motion and pressure, and the sting of a great heat throughout his body—almost—and then he was a gas, incandescent, streaming out north and northeast at two hundred miles an hour. At precisely that moment, verification of his fate came from an automated geology laboratory located on the south shore of Spirit Lake, about a thousand yards east of Truman's lodge. A geodimeter had gone off the air in the middle of a broadcast, melting and squealing as it went.

Beyond the lodge, more than eight miles away, all the trees were snapped down as though they'd been nothing more than matchsticks standing on end. But though they had been felled as easily as matchsticks, they were more like compass needles: each of them, as they collapsed in rows by the tens of thousands—by the *millions*—recorded a phenomenon no one had seen before; each pointed the way that the water-dense wave of air had gone. The collective whole painted a picture of myriad channels, eddies and unexpected course changes within the death cloud. Here, the blowdowns diverged in two directions from an ash-choked streambed, looking like parted hair. There, the cloud had weaved and bumped on the contours of the land. It was capricious, like a tsunami.

Even its heat was curiously inconsistent. Almost seven miles north of Truman's lodge, campers Bruce Nelson and Sue Ruff somehow survived the temperatures and the velocities, and the toppling of 150-foot Douglas firs over their campsite, while only a thousand feet away from the Nelson camp, trees were seared, other campers burned to death, and a remote laboratory registered 750°F. The survival of John and Christy Killian was even more surprising. They were hiking near the north shore of Spirit Lake, five miles closer to Truman's lodge, in an area where trees were practically vaporized as they fell. Temperatures there reached 1200°F.

Apparently there were voids within the death cloud, random and rare eddies filled with cool air. It's anybody's guess where the cold spots came from, or what features of the landscape kept them stationary, in the rushing air and dust, long enough to spare the Nelson/Ruff and Killian parties; but wherever the voids appeared, they seem to have behaved like protective bubbles.

During that same spring season of 1980, excavators at Pompeii's sister city Herculaneum were making equally unexpected discoveries. The ruins of the city had preserved an independent record of almost identical events.

Skeletons were found, nineteen hundred years old and perfectly intact. A soldier had been slammed face down by an invisible fist that hurled roof tiles, stones and a middle-aged woman along a street of row houses. The woman impacted like a missile near the fallen soldier. Her bones were shattered in two hundred places. The soldier was pushed down so hard that he never moved again. His sword, and a purse containing three gold coins (one bearing a portrait of the emperor Nero), are still strapped to his waist.

The force of the fist is a puzzle. It varied greatly. On one block, it left hanging gardens undisturbed. On another, a glass windowpane exploded indoors, and the air, driving and sucking as it jetted in, overturned a marble basin and crashed it against the far wall. The hot ash, too, played strange tricks.

"For all its force and massive volume and high temperature," says archaeologist Joseph Jay Deiss, "the cloud was inherently unpredictable. With modern scientific techniques it has become possible to estimate the peak temperature of the ash. Using infrared spectrometry, the structural and chemical changes that occurred during the thermal degradation of wood can be determined." It appears that wood in Herculaneum was heated to 752°F in some places, yet in others, the ash flow was hot enough merely to scorch cloth and bread. In one house, on the same city block where temperatures of 752°F were recorded, wax seals remained unmelted, still holding together paper scrolls in which the story of a Roman lawsuit could be read.

The burial of Herculaneum differed from the burials of Pompeii and Thera, and from every other volcanic tomb in the world. At Herculaneum, the flecks of hot ash fused together, and consolidated into a protective rock mass impervious to water, sealing the city inside as if it were embedded in Cretaceous amber.

Herculaneum differed in one other important aspect. The death cloud from Vesuvius, according to an eyewitness account that still survives, was barely twenty miles wide. Thera's black shroud spread over much of the Aegean, and appears to have had lethal effects even hundreds of miles away, in such places as Turkey and Egypt. Vesuvius and St. Helens combined added up to only one cubic mile of airborne debris. Thera ejected almost thirty times as much.

If pockets of glassy volcanic ash in sea sediment are any indication, the Theran death cloud surged over the eastern half of Crete, covering both the north and south shores. On easternmost Crete, more than seventy miles away from ground zero, the great palace of Zakros fell amid flames and ashy deposits. Stone slabs were slammed horizontally across the ground in a manner originally attributed to an earthquake, but all the stones seem to have toppled in the same direction, as if pushed over by a great wind. Like Herculaneum, Zakros perished so quickly that people did not have time to flee with household objects. All the implements of Minoan life were left behind: gold rings, razors, tweezers and rare perfumes. At the same time, Phaistos, second in size only to Knossos, was utterly carbonized on the southeast coast. Knossos itself seems to have been spared the absolute destruction visited upon Phaistos, despite the fact that Phaistos was located twenty miles farther from Thera, on the far side of an eight-thousand-foot mountain.

Such inconsistencies are a puzzle to archaeologists and historians, but new discoveries at Herculaneum, and the evidence of Mount St. Helens, suggest that they are to be expected, that wave fronts of hot gas and volcanic ash, as they surge over the ground, may dance and weave like snakes, and are by their very nature inconsistent.

When a volcano explodes, the rock it pulverizes moves immediately in two directions, depending on how large the pulverized pieces are. The larger, heavier fragments, ranging down in size to grains of sand, fall to earth, each trailing a little slipstream of air behind it, and the collective whole creates a mighty downblast, which, blocked by the earth itself, flows out laterally, hugging the ground. Surging forward at more than a hundred miles an hour, each particle shedding heat as it moves, this is the death cloud, sometimes known as the "Pelean phase," after the eruption of Mount Pelée on Martinique in 1902, in which a ground surge swept through the city of St. Pierre, killing almost all of its thirty thousand inhabitants.

The smaller, microscopic particles are influenced more by the air into which they shed their heat than by the pull of gravity. Hot air rises, hoisting the fine ash into the stratosphere. This is the "Plinian phase," named after the young scientist who first described it, from a vantage point fifteen miles west of Herculaneum.

"A remarkable phenomenon," Pliny the younger wrote to the Roman historian Caecilius Tacitus, in the only account of the eruption to survive from antiquity.

The cloud was rising. . . . In appearance and shape it was like a tree. . . . Like an immense tree trunk it was projected into the air, and opened out with branches [the Plinian phase].

The earth shocks . . . became so violent that it seemed the world was not only being shaken, but turned upside down. Whether from courage or inexperience . . . I called for a volume of Titus Livius and began to read, and even continued my notations from it, as if nothing were the matter.

Though it was the first hour of the day, the light appeared to us still faint and uncertain. And though we were in an open place, it was narrow, and the buildings around us were so unsettled that the collapse of walls seemed a certainty. We decided to get out of town to escape this menace. The panic-stricken crowds followed us, in response to that instinct of fear which causes people to follow where others lead. In a long close tide they harassed and jostled us. When we were clear of the houses we stopped. . . . The sea appeared to have shrunk, as if withdrawn by the tremors of the earth. In any event the shore had widened, and many sea creatures were beached on the sand. In the other direction loomed a horrible black cloud ripped by sudden bursts of fire, writhing snakelike and revealing sudden flashes larger than lightning [the death cloud, or Pelean phase].

Soon after, the cloud began to descend upon the earth and cover the sea. It had already surrounded and covered Caprae, and blotted out Cape Misenum. My mother now began to beg, urge and command me to escape as best I could. A young man could do it [the cloud by now had thinned, slowed, and shed much of its heat, and might even be outrun by a man]; she, burdened with age, would die easy if only she had not caused my death. I replied that I would not be saved without her. Taking her by the hand, I hurried her along. She complied reluctantly, and not without self-reproach for hindering me.

I turned around. Behind us an ominous thick smoke, spreading over the earth like a flood, followed us. 'Let's go into the fields while we can still see the way,' I told my mother—for I was afraid that we might be crushed by the mob on the road in the midst of the darkness. We had scarcely agreed when we were enveloped in night—not a moonless night or one dimmed by cloud, but the darkness of a sealed room without lights. To be heard were only the shrill cries of women, the wailing of children, the shouting of men. . . . Many lifted up their hands to the gods, but a great number believed there were no gods, and that this night was to be the world's last, eternal one.

Pliny's letter is the earliest known record of the step-by-step behavior pattern of an explosive volcano. Its last line is also telling. Vesuvius was a fairly localized event, yet even before the story had been passed down to a single generation, survivors had begun to view it, and tell it, as the end of all the known world. It is easy to see how an even larger story, the Theran one, can come down through time swollen to continental proportions, if not to the end of the world itself.

Pliny was fifteen miles away from the downblasted, burned and buried cities in the east. When he eventually journeyed there, he found everything changed. Herculaneum and Pompeii were covered by a thick layer of ash, like an unnaturally deep snowfall—only the tops of the tallest buildings poked through. We have no eyewitness accounts of how the "horrible black cloud . . . writhing snakelike" behaved within the cities. Such close-up views would have to await Mount Pelée and the destruction of St. Pierre.

SPRING, A.D. 1902

On the exceptionally clear morning of Thursday, May 8, 1902, Gaston Landes, St. Pierre's leading geologist, stood beside a lily pond in the city's ash-shrouded botanical gardens. "Take away the ash and the sulphur in the air, and the actual damage caused by Pelée would so far appear to be small," he'd told Governor Louis Mouttet. "It's just an eruption. Don't worry about it."

For almost two weeks now, the mountain had been stirring to life, quaking and booming and sending up flurries of warm ash. It could not have happened at a worse time: it was an election week. The governor had decided that "at all costs, the population of St. Pierre must stay in the city."

Assured by Landes that the volcano presented no threat to human life, the governor went so far as to suggest that the sulphur fumes were good for people's health, and then posted troops to block Le Trace, the only road out of St. Pierre. He continued to post them even after clouds of sulphur and ash had choked off the first lives.

Landes, a professor of natural science, knew of two kinds of volcanic activity that could threaten the city: mudslides and lava flows. Even if lava, or mud, or both did flow, he reasoned it would have to cross three valleys, each more than a hundred feet deep, to reach St. Pierre. "The valleys are natural dams," Landes said; and the governor took him at his word.

There was a third kind of eruption—to which Pelée would give its name—that Landes did not know about as he stood in the garden brushing ash from his shoulders. Overhead, the twin towers of the cathedral glistened in the morning sun. Then from the nearer of them came the sound of bells summoning the faithful to Ascension Day mass. It was 8:00 A.M. In two minutes Pelée would begin to teach. It was a lesson Landes would not live long enough to profit from.

A half mile east of the gardens, nineteen-year-old Auguste Ciparis crouched in a tiny cell on the basement level of the local prison. In a barroom brawl he'd killed a man, and for that he was sent to death row. Now, for a second time, and in a way Ciparis could never have anticipated, the fact that he'd been faster with a knife than his opponent meant the difference between life and death. As it turned out, death row was one of the safest places in town.

In the harbor district, Léon Compère-Léandre had returned home after a few days' absence to find all the rooms occupied by refugees. The governor's office had announced that—far from evacuating the city—people in outlying areas should be brought into the safety of St. Pierre. Two thousand had come, but there was no place for them to stay, so they began breaking into any home that appeared to be vacant. Léon warned the intruders that if they did not leave, he would go to the town hall to seek help in evicting them. They pleaded with him, even offering money.

"I am not an unreasonable man!" he shouted. "It's just that you cannot allow people to take over your home."

He turned and was heading out the door when a woman changed his mind—and saved his life. She held a caged bird in her hand. "Please, mister," she begged. "If I have to go, please let me leave the bird with you for safekeeping."

He almost wept, for no reason that he was consciously aware of. "OK," he relented. "You can all stay . . . for a little while."

He then climbed down to the wood cellar, the only room that had not been commandeered, and spread a blanket down on a pile of stacking.

Out in the harbor, the crew of the luxury liner *Roraima* was clearing away a layer of white volcanic ash that lay fully a quarter-inch deep over everything. "When the captain and I came off the bridge," recalled First Officer Ellery Scott, "our uniforms were covered with it. Passengers and crew were gathering up the sand and ashes to keep as mementos. Some had put it in envelopes, others in tin tobacco boxes, and I remember a tall black man giving me a cigar box filled with it, which I took, little thinking what a plenty I would have before I made home again."

At 8:02, as a knot of about forty people stood on a hill overlooking the city, preparing to descend for the morning mass, someone in the telegraph office tapped out ALLEZ—the last word anyone heard from St. Pierre.

Then, where the mountain had been, a ball of glowing red dust appeared. It swelled, tripling its size in a second, then tripling again. As its edges surged through cool air, the ball faded from red to black. Four miles away, amid the hillside knot of people, Fernand Clerc stood with his family. They held each other close, waiting for death and wanting to die together. The thing held Fernand spellbound—there was something dreadfully beautiful about it—all roiling black hell and expanding dimensions. For a moment it seemed fixed to the mountain; but then part of the cloud broke off and, with astonishing speed, advanced along the ground toward the city.

The soil and rocks rumbled, but there was as yet no sound in the air above. Shock waves are transmitted more rapidly through the solid earth, so the Clercs saw the top of the mountain burst apart, and felt the release of some twenty or thirty kilotons beneath their feet, eight seconds before they actually heard it. And when they did hear it, it arrived as a concussion of trapped air strong enough to knock them down.

"Above, a column of dust mushroomed and blotted out the

sun," observed Mrs. Clerc. "Below us, we saw a sea of fire cutting through the billowing black smoke and washing over the three valleys that were supposed to protect the city from Pelée. And incredibly, within these high-walled valleys, there existed havens of refuge where trees and bushes survived in the lee of one massif—the plane of destruction passing harmlessly overhead. The cloud tumbled over and over. One moment it would clutch at the ground, the next it would rise perhaps a hundred feet before falling back to the earth again. It seemed to be a living thing. It leapt over the botanical gardens, and in places even doubled back on itself, traveling the way it had come."

People on Le Trace saw the thing coming, and started to run as fast as they could toward the Clercs. At a distance they looked like swarms of ants. The black death rolled impartially over them and erased them from view, except for a man on horseback who managed to keep ahead of the cloud, actually outpaced it on an uphill run and survived. Behind him, the prison district was swallowed whole. The thing snaked out toward the harbor and the *Roraima*, eating up the governor's mansion and Léon's house with equal ease. For a moment the *Roraima*'s stern jutted out of the blackness, then it too was swallowed, as a raindrop is swallowed by the ocean. Only the towers of the cathedral seemed a safe, untouched haven, standing clean and bright above the death cloud; but they only seemed so, for soon they sagged and glided down into the wash.

On the *Roraima*, Ellery Scott had been standing on deck with a fellow crewman who, at 8:02 A.M., had commented on the "peaceful sight of St. Pierre," and gone below deck to retrieve his camera. Scott never did see him again, for seconds later came "darkness blacker than night, and as the awful thing struck the water, it just rolled along, setting fire to the shore and ships. . . . The masts, smokestack, rigging—all were sheared clean off to two feet above the deck, perfectly clean, without a jagged edge, just like a clay pipestem struck with a big stick."

The masts and smokestack had been snapped by nothing more

than a surge of air made heavy with rock. In the city, a hundred yards beyond Scott's position, trees had been snapped down in much the same manner—usually in rows, all pointing in the same direction—exactly like trees washed over by the St. Helens death cloud. There were other, equally important similarities to the St. Helens eruption, and to what excavations at Herculaneum and Pompeii are slowly revealing about ground surges.

"There was something strangely inconsistent about the cloud," recalled Scott. "I was standing out in open air, yet other people, in what seemed to be more protected quarters, were dead within seconds. . . . The darkness was something appalling. It enveloped everything, and was only broken by burning clouds of consuming gas which gave bursts of light out of the black. . . . A red glow approached, then veered away to starboard.* The *Roraima* took fire in several places simultaneously. . . . The dining saloon and the stern blazed up at once, and yet stores of kerosene on the bow failed to ignite . . . while on a ship practically right next to us, barrels had exploded into walls of flaming rum. The *Grappler* did not even have time to burn before she capsized and sank. Not one of her crew survived. They were the fortunate ones.

"For myself, when I saw the thing coming, I snatched a tarpaulin cover off one of the ventilators and jammed it down over my head and neck, looking out through the opening. This saved me much, but even so the inside of my mouth burned. I heard my hair sizzle. . . . Then, all at once, the ash was settling a little, letting some of the sunlight through . . . twilight . . . it was possible to breathe, and see . . . to see too well, too much. All

---

* What Scott has described here is probably the passage, very near to him, of hot spots within which the ash still glowed red. By the time the death cloud reached the *Roraima*, dust particles had already shed much of their heat to the air they passed through, having faded from white hot to red and finally to "blacker than night." The air itself was now hot enough to scald, but there were evidently hotter things moving busily to and fro within the cloud. The pockets of insulated, still-glowing ash were essentially roving fireballs, and Ellery Scott was the first man ever to see such things and live to describe them. This may account for the survival of Mount St. Helens' Killian party in a zone of vaporized wood. It seems a barrage of 1200°F fireballs just happened to miss them.

around were sailors and passengers, men, women and children, burned and dying, crying aloud for water. . . . There were only a few of us really able-bodied. . . . Gradually we collected the survivors and laid them on deck forward, near number one hatch. All of them cried for water. But many of them could not drink at all. The cloud had burned their mouths and throats and even the linings of their stomachs, so terribly that in many cases the passage of the throat was almost entirely closed. When we put the water into their mouths it stayed there and almost choked them, and we had to turn them over to get the water out, and still they would implore us for more. . . . The fire did not seem to penetrate clothing, but wherever the flesh was exposed, it burned mercilessly.

"The women passengers behaved very well, though they were all terribly burned . . . and they knew that the ship was burning and might be sinking. They knew that all the lifeboats had been smashed and that we'd have to build a raft and be prepared to shove off if worse came to worst before some other means of assistance came to us. . . . One very big woman, for all her burns and scalds, kept singing hymns. Between the verses her cry was the same as the cry of everybody, 'Give me water, water.' As soon as she got a drink, she seemed to revive, and then she would begin her singing again. The last hymn she sang was 'Nearer, My God to Thee,' and then she died where she was sitting.*

---

* This particular incident resounded through both the mythology and slang of the twentieth century. In St. Pierre, during the weeks leading up to the May 8 eruption, the governor had assured his constituents after every volcanic display that Pelée's activity was on the wane, that it was over. In Pelée's aftermath, a then obscure expression among wealthy opera audiences—"It's not over till the fat lady sings"—was brought out into the mainstream and endures to this day. The unidentified lady's singing, which had a calming, possibly life-saving effect on survivors, immediately became legend, but, much like the story of Thera and Crete received by Solon, would reach us misunderstood and distorted. Ten years later, American tabloids, seeking to sell newspapers by dramatizing an already dramatic event, would borrow the *Roraima's* "Nearer, My God to Thee" story (which had moved the world to tears in 1902), and transfer it to the *Titanic's* band. During the 1950s, two films would portray the band playing the hymn as the *Titanic* sagged into the Atlantic, and from that moment a popular

"Here and there through the smoke, ruins showed. In three minutes, all had been demolished. St. Pierre was a dead city, filled with dead people."

In his cell, Auguste Ciparis stirred uneasily, trying to understand why he was not dead. No one and nothing else moved. The air smelled of wood cinders and cooked flesh—and he realized that part of that smell was his own skin. Through his tiny window he'd seen the dust wave coming, eating up the sky and coming fast. Fearing that it would choke him, he had urinated in his shirt, covered his face with it and breathed through it. The cloud shot through the upper stories of the prison, turning them into a brick pile. A man standing in the yard outside was caught in a particularly hot stream of dust. It turned him into a carbonized statue where he stood. As he fell, he shattered like charcoal. Ciparis did not see this; bricks collapsed on top of him, molded around him, and saved him from swift death. He had tried to free himself from the bricks, but was knocked back by the blast of superheated air and dust jetting through the cell window. Suddenly it stopped, and an unbelievable silence fell over the place, broken only by the dull roar of flames, and by a woman's voice carrying out over the harbor. Ciparis thought he recognized a Welsh hymn, but it ceased abruptly.

Less than a mile away, Léon Compère-Léandre climbed out of his cellar into the fringe of hell. The whole top of his house had been broken off at the middle and hauled away, yet his waistcoat was still hanging on the wall exactly as he'd left it. As he puzzled over this, the coat burst spontaneously into flames. The woman who had offered him the bird was burning in a corner. Her pet was carbon. Léon moved to help her, but she stood up and ran from the house, trailing sheets of fire. She ran perhaps a half

---

myth became historic tradition. This is no more a historic fact than the story about George Washington throwing a silver dollar across the Delaware River (which is true only in the sense that the dollar used to go a lot further). For trivia enthusiasts, the last piece played by the *Titanic*'s band was *Songuede Autumn*, a popular 1912 waltz.

block, in the direction of the cathedral. Then she fell and did not rise.

Most of the refugees had been stripped naked by the cloud. Their hair, too, was gone. In a bed he found a man dressed in what he first thought to be a red sweater. The man was dead, but still moving. As Léon watched, the body inflated rapidly, as if somebody were pumping compressed air into it. When finally, inevitably, the abdomen broke open, something yellow and ropy squeezed out and turned red.

Léon stepped away, gagging, and started to run. Bolting out the door—or, rather, the place where the door had been—he slipped on something soft, recovered and stumbled against a kneeling man whose body had already burst open. "The head was scalped, burned, the eyes gone, the lips formless, surrounding something black which turned out to be a tongue made of charcoal."

Around him, the city looked as if it had been in ruins for a thousand years. Beyond his house, he saw no sign of St. Pierre's thirty thousand inhabitants. Then he looked again and noticed bundles of dust-covered clothing. They lay thickly near the demolished cathedral. The cloud must have caught them rushing out into the street. One of the twisted shapes stood up and began moving toward him. Léon wondered why the man—or was it a woman? he could not tell—was carrying his shirttails in his hands. Then he realized that the man (woman?) was carrying his own skin.

Léon turned and ran, yelling something incomprehensible. This time, nothing would stop his flight, not even the prickling sensation that grew in strength each passing second, and made him aware, for the first time, that he, too, was burned. His legs, arms and chest were blistered, but apart from those wounds he did not appear to be hurt at all. His blisters would heal. The death cloud had spared him, physically.

St. Pierre was a cityscape that made Dante's hell look shamefully mild. And all the horrors hatching out—the blackened things, the charcoal people, the dead and the still-moving dead—all of these had required just ten thousand cubic yards of vaporized rock dusted less than half an inch deep over eight square miles of the city. Survivors' accounts of the Pelée death cloud provide only the slightest glimpse of how the eastern half of Crete must have suffered in the aftermath of Thera.

Thera blew a thirty-cubic-mile hole in the earth. Thera was fifteen million times worse than Pelée.

WINTER, A.D. 1989

Doumas tells me that all the finest bronze spears, and all the gold and silver that must have been present in the Therans' homes, are curiously absent. Barely a trace of them has ever been found.

The fact that everything of value seems to have been removed from the buildings suggests that the Therans were careful, and heeded the warning signals of an awakening giant. They'd evacuated days, weeks, years and possibly decades ahead of the final upheaval, but in the end it did them no good. They could not have anticipated the tidal waves, and the enormous dust cloud whose mass alone behaved as an insulating force, probably over many tens of miles, keeping in the heat, preserving random hot spots that, as they weaved and bumped over the earth, must have been promptly lethal even a hundred miles away. The hot and cold spots may account for one of the paradoxes of widespread fire destruction on Crete. Carbon dating and traces of volcanic debris in Cretan ruins suggest that the burnings were approximately synchronous with the deposition of a Theran ash layer over much of the eastern Mediterranean seabed. Yet the fact that Knossos was largely spared, while Zakros apparently burned instantly to the ground, that even at closer quarters, mansions fell in smoke and flames while ordinary houses one thousand feet away stood intact—and vice versa—has caused archaeologists to wonder if all the damage could not have been caused by wild tribes of invading Greeks. Volcanic death clouds do not act with such inconsistency, it has been reasoned; crazy humans do. The records of destruction and survival at St. Helens and St. Pierre suggest otherwise. The recent excavations at Herculaneum (where wood was carbonized by hot volcanic dust in one room, while in another room buried under the same surge, amber and wax did not even melt), suggest that death clouds can be every bit as inconsistent as human beings.

Even where the cloud was not lethally hot, it had many ways of killing. Systematic deep-sea coring by the research vessels *Trident* and *Atlantis II* indicates that the ashfall over eastern Crete accumulated to a depth of approximately two inches.* A couple of inches may not sound very dramatic; but when I examined photographs of the ash-covered bodies at St. Pierre, and measured the thickness of the cover (a task made possible by swollen flesh, which had opened up cracks and revealed measurable cross-sections through the ash layer), I was surprised to learn that less than a half inch had fallen upon the city. This is only about twice the depth of Theran ash now being found in Egypt—although it is probably safe to bet that the Theran cloud had shed most of its heat by the time it reached the Nile.

It is also important to note that the Mount Pelée death cloud, small though it was, killed vegetation on the island even where it fell cold enough not to be lethal, even where it covered leaves only a sixteenth of an inch deep. Crops all over the island blighted and yellowed, causing famine and economic collapse.

In 1815, when Tambora let loose with a Thera-magnitude explosion, a one-inch fall of cool ash, 250 miles away in Makassar, blotted out the sun for two days, choked all the birds to death, blighted crops and killed most of the fish in the lakes. A week

---

* The ash layer in the deep-sea mud immediately north and south of Crete is two inches thick, suggesting that the island itself must have been covered at least as deep. On a mountainous, rain-washed island like Crete, land-based core samples are not likely to give an accurate history of the sedimentary environment. The best places to sample Crete's history are therefore in the relatively undisturbed deep waters that surround the island.

later, rats, which must have become desperately hungry, were seen hunting animals larger than themselves. On western Sumbawa, 125 miles from Tambora, the ash accumulated more than a foot deep. During the weeks that followed, thirty-eight thousand survivors died from starvation and disease.

More than seventy miles east of Thera, directly in the path of the cloud, Southern Turkey and the islands of Kos, Rhodes and Cyprus received more than a foot of ash. In those places, the cloud, even if it had shed all of its heat (which it probably had not), would have suffocated almost everyone caught outdoors.

St. Helens gives us some idea, a very small and vague idea, of the manifestations east of Thera. Eighty-five miles downwind of Mount St. Helens, the town of Yakima was enveloped in darkness at midday. Yakima received only about a half inch of warm ash. The ash itself was comprised primarily of microscopic particles of glass, some of which stuck to the skin like microshards from a broken soda bottle.

"The cloud was a mass of cutting edges," observed one eyewitness. "This black, black cloud, of a blackness I have never seen, came over. . . . No headlights could cut through it. . . . When we first saw it coming, we expected the most horrendous thunderstorm of our lives. We were right—lightning struck everywhere, jarred the ground—but it heralded ash instead of rain."

Within the cloud, birds and insects fell dead to the ground. People who did not mask their mouths and noses with cloth were coughing blood by the time they staggered into the hospitals.

Cameras photographed Mount St. Helens in color and 3-D from every conceivable distance and angle. Civilizations have seen the disappearance of cities and the formation of tidal waves two hundred feet tall. Volcanic mudslides have washed away whole districts, and new islands have formed before our very eyes; but human experience can scarcely furnish criteria for occurrences in and around the Mediterranean during Thera's last hours. When finally the cloud had passed, thousands of bodies lay bleeding under the starlight on Kos and the other islands east of Thera. The towns, those that were above the tsunamis, seemed to be sleeping peacefully under drifts of fresh-fallen snow—but it was not snow, and not even a spider stirred upon it.

On eastern Crete, where the ash fell only two inches deep, it must have smothered some of the plant life, livestock and people. An important point to bear in mind is that the cloud need only have demoralized the Minoans, need only have brought about their economic collapse—something comparable to the stock market crash of 1929, or the shaky state of our present global economy—to weaken Minoan Crete and eventually render her people subservient to mainland Greeks. And the death cloud was not the only weakening force: there were, of course, tsunamis, at least one of which left fossil traces of itself.

Thera was apparently spewing uncountable millions of tons of pumice stone during the weeks or months leading up to the final explosion. This glassy froth actually floats on the sea, and mats of this material, chemically traceable to Thera's second millennium B.C. eruption, have been found ninety feet above sea level on the island of Cyprus, in quantities large enough to have made mining profitable. The question that troubles the miners is, how did it get there?

We know from satellite observations of a 1976 eruption in northern Tonga that huge rafts of pumice can be exuded by volcanoes. The Tonga raft was, at one point, almost as large as the state of Rhode Island, and satellites were able to track it for three months. Such rafts must have formed during the weeks leading up to Thera's final upheaval. One of them evidently drifted to Cyprus and had begun to bump up against its shore when Thera exploded. The tsunami then lifted the raft, which explains the high altitude pumice deposit. In the fields below the Theran pumice, an ancient Cypriot town seems to have been jostled about. Large stones were broken in half, and the upper halves were moved hundreds of feet, just like the stones Spyridon Marinatos first encountered in ruins on the Thera-facing shore of Crete.

From the eastward spread of the ash cloud, we know that a gale must have been blowing in the general direction of Turkey when the island exploded. Oceanographic core drillings have revealed ash deposits two feet deep on the sea floor near the Turkish and Cypriot coasts. It is anyone's guess what the ashfall did to tuna populations, the main source of protein in the eastern Mediterranean. Three hundred miles south of Cyprus, the Nile delta was out of the main line of fire: nevertheless, recent excavations suggest that at least a quarter inch of dust fell out of the cloud as it rolled over Egypt. The dust was rich in sulfur, and some of it appears to have fallen as acid rain. Doubtless it blacked out the sun for several days, and as it circled the globe, the next spring must have been much like the spring of 1816, "the year without a summer," that followed the explosive reawakening of Tambora in Indonesia. There were no harvests in New England that year. Megatons of ultrafine dust had been hoisted fifty miles high into the stratosphere, where it shaded out some of the sun's radiation, absorbing its heat long before it reached the ground. As June and August snowstorms swept across New York, few people could draw consolation from the strange beauty of a blood red moon, or from the most splendid sunsets the world had seen in more than thirty-four hundred years.

Growth rings in California bristlecone pines—"the trees that live forever"—display frost scars from the false Tambora winter. The rings are laid down annually, and one need only count back year-by-year to date the scars. We know from the pines that there was also frost in California during the summer of 1627 B.C.* At around that same time, a thin layer of dirty, acidic snow was deposited on the Arctic ice cap. Plato did not write about the long Theran night in his tale of Atlantis, but the Egyptians did: ancient texts describe a thick, choking dust cloud that devoured the sun, and tell of towns being swallowed by waves.

---

* Note that the time scale used in this book requires that we continue back through a year "0" inserted at the A.D./B.C. transition.

"It is inconceivable what happened to the land," begins one Egyptian text.*

> The land—to its whole extent confusion and terrible noise. . . . For nine days there was no exit from the palace and no one could see the face of his fellow. . . . Towns were destroyed by mighty tides. . . . Upper Egypt suffered devastation . . . blood everywhere . . . pestilence throughout the country. . . . No one really sails north to Byblos today.† What shall we do for cedar for our mummies? Priests were buried with their produce [objects of foreign trade], and nobles were embalmed with the oil thereof as far away as Keftiu [Crete], but men of Keftiu come no longer. Gold is lacking. . . . How important it now seems when the oasis people come carrying their festival produce. . . . The Sun is covered and does not shine to the sight of men. Life is no longer possible when the Sun is concealed behind the clouds. Ra has turned his face from mankind. If only it would shine even for one hour! No one knows when it is midday. One's shadow is not discernible. The Sun in the heavens resembles the moon.

It could all be a description of a very bad dust storm blowing in from the desert—but the Egyptians were used to such things. This particular storm was recorded by scribes because it differed from anything within their experience. And what are we to make of "terrible noise, pestilence, blood everywhere, devastation," and the interruption of shipping, in particular from the direction of

---

* The Ipuwer papyrus is believed to be a Middle Kingdom document dating from a tumultuous period during which five dynasties, three Egyptian and two foreign, vied for power, sometimes ruling simultaneously. Most scholars place the Ipuwer papyrus in the Thirteenth Dynasty, traditionally dated about 1600 B.C., but it is important to note that insufficient knowledge makes dating everything except the Thera ash layer (which can be dated in Egypt because its effects were recorded half a world away in tree rings and glaciers) very difficult. Age uncertainties of one hundred and even two hundred years seem likely. In fact, future archaeologists will probably be forced to reassign Egyptian dynasties based upon whether artifacts are found above or below the ash layer, and how far from it.

† Byblos was a port of Lebanon, a country famous in ancient times for its cedars. The port also gave its name to a fine papyrus it exported, and this word for a papyrus scroll lives on in "Bible."

Crete and Thera? Doumas and I have been toying with the idea that the scribes have recorded the passing over of the Theran death cloud, and that this may account for some of the plagues of Egypt.

Egypt's Ipuwer papyrus and the Old Testament seem to be describing the same event. When we review what we already know about the Thera eruption, and compare the archaeological and geologic evidence with the more recent and therefore better preserved, more detailed accounts of Tambora, Pelée and St. Helens, a story begins to emerge, with seemingly disparate events becoming as intimately connected as pieces of a spider's web.

Item. "Blood everywhere," laments the Ipuwer papyrus. If one speculates on this statement, "blood everywhere" could be a reference to people coughing up blood after inhaling the ash of Thera, or to the Nile River, polluted with acidy, sulfuric ash, filled with rotting fish and transformed, in effect, from the bringer of life to the river of death. The Old Testament claims that "the fish of the river died, and the river itself became so polluted that the Egyptians could not drink its water. There was blood throughout the land of Egypt." This account of the first plague is an apt description of what the ash of Tambora, Pelée and St. Helens did to lakes and rivers in the paths of their death clouds: fish died, and water turned from glass-clear to an acidic murk, sometimes the color of drying blood.

Item. The Old Testament makes references to plagues of frogs, gnats, flies and locusts. There are even descriptions of snakes and other vermin "coming up from the Nile—into your palace and into your bedroom and onto your bed. . . . The dust of the earth was turned into gnats throughout the land of Egypt. . . . The frogs in the houses and courtyards died off. Heaps and heaps of them were gathered up, and there was a stench in the land . . . and all the livestock of the Egyptians died. . . ." Egypt's Ipuwer papyrus provides glimpses of similar events, as do accounts from Mount Pelée in 1902. Ashfalls and hunger weakened the inhabitants of Martinique, allowing disease to break out and spread unchecked. Birds and livestock choked to death on the dust far more easily than humans, and everywhere there was the unsettling stench of decaying animals. Those creatures that were not killed outright by the ash were "made bold by hunger," and behaved in strange ways, reenacting scenes straight out of Exodus and the plagues of Egypt. Ants were driven by Pelée from the banks of Martinique's Blanche River. With them came centipedes more than a foot long, baring pincers thick and powerful enough to pierce shoe leather. They swarmed into barns, up the walls of parlors, into bedding and night clothes. Rats attacked children on the streets, and fer-de-lances, the deadliest of the island's snakes, descended upon a single town in a great swarm. Within thirty minutes they had killed at least as many children and twice as many horses, dogs and pigs. A sow was seen trying to defend her litter from three of the giant brown snakes. They sprang at her, then destroyed her litter "in a series of tongue flicks."

In the shadow of Mount Pelée, one eyewitness account made it easier to understand why ancient Egyptians worshiped their cats.

> One of the most remarkable battles I had ever seen: the cats, whether through hunger or bravery, advanced boldly on the serpents until they were just outside striking range. Then the cats would feint, teasing, startling, or trying to draw the serpents to strike first. And when they struck, heads hissing from coiled bodies, the cats pawed them aside, mangling the heads of the fer-de-lances. Again the snakes would strike—and again the cats would sweep the heads aside, inflicting further damage. Blinded, scaled skin deeply torn, the fer-de-lances were stunned. Then, the cats leapt on them, nailing their heads to the ground while keen white teeth severed their vertebrae. In an hour, over a hundred snakes had been bitten to death.

Item. The Old Testament tells us of hail mixed with flames raining from heaven, and peals of thunder, and lightning flashing down to earth. Such things were witnessed as the Mount St.

Helens cloud (in its Plinian phase) passed over Yakima, lancing down lightning and warm ash instead of rain, and bringing with it "a darkness that was more than darkness." The Ipuwer papyrus makes reference to "darkness . . . confusion and terrible noise throughout the land."

"Then," says the Old Testament, "the Lord said to Moses, 'Stretch out your hand toward the sky, that throughout the land of Egypt there may be such intense darkness that one can feel it . . .' and there was intense darkness over the land of Egypt. Men could not see one another, nor could they move from where they were for three days."

Again, Egypt's Ipuwer papyrus recounts a familiar story: "The land—to its whole extent . . . darkness in Egypt . . . for nine days there was no exit from the palace and no one could see the face of his fellow. . . . One's shadow is not discernible."

The language is so hauntingly similar that one can easily believe that the Exodus account and the Ipuwer account were derived from the same scrolls.

Item. After the Exodus, there arose among the Hebrews a historic tradition of associating God with a fiery cloud. It is possible that the tradition survives as a distorted and poorly understood memory of Theran ash storms. In the Book of Numbers, the Lost Ark of the Covenant (the golden chest in which the Hebrews carried the Ten Commandments) was often overshadowed by God's cloud, which struck out and consumed enemies, and on occasion even set part of the Hebrew camp afire: "During the day the Dwelling [of the Ark] was covered by the cloud, which at night had the appearance of fire . . . and when they set out from camp, the cloud of the Lord was over them by day. Whenever the Ark set out, Moses would say, 'Arise, O Lord, that your enemies may be scattered and those who hate you may flee before you.' And when it came to rest, he would say, 'Return, O Lord, you who ride among the clouds, to the troops of Israel.' "

When finally the troops entered Canaan, carrying the Ark before them, war broke out almost immediately between the Hebrews and the people they found there. Among those people were the Philistines, whom the Bible tells us came from Caphtor (Crete). Can it be that the Philistines (Cretan Minoans?) and the armies of Hebrew slaves, having escaped from (or been chased out of) famine-stricken Egypt, were actually two populations of refugees created, in different ways, by the same volcanic catastrophe? Can it be that the present-day conflict between the Palestinians and the Israelis has as its roots Thera and the origin of the Atlantis legend?

Item. Jeremiah, who lived about 610 B.C., defined the Philistines as a remnant from the coasts of Crete, and is said to have prophesized an angry God's vengeance upon these "seashore enemies of the children of Israel." Like many biblical prophesies, the Book of Jeremiah (47:1–4) may project earlier events into the future. The Philistines' tribulations are hauntingly reminiscent of the Theran tsunamis of 1628 B.C., which the tribes of Canaan are likely to have associated with the arrival of a lost, seafaring tribe from the northwest. "Behold," says Jeremiah. "Waters are rising from the north, a torrent in flood, it shall flood the land and all that is in it, the cities and their people. . . . Fathers turn not to save their children; their hands fall helpless. . . . Yes, the Lord is destroying the Philistines, the remnant from the coasts of Caphtor."

The Bible, archaeology and geology are beginning to converge and agree on at least one point: that the Exodus really did happen, and that a cloud from Thera might have been the tool that struck fear into an Egyptian pharaoh, causing him to let Moses and his people go. If this was indeed the case, then an important theological puzzle emerges.

When I mentioned a possible Thera-Exodus connection to a friend of mine, a Jesuit priest who likes to send me off on every expedition with a special blessing, he became a little shaken. When you come right down to it, what we are talking about is the annihilation of a once fertile and beautiful island, and the peaceful, seafaring Minoan culture that thrived on and around it.

"I can accept the end of Thera as a random accident of nature," my friend says. "But if we are to believe it was God's wisdom at work—there were so many other ways that God could have sent plagues upon Egypt, so many ways to darken the sky, cause the fish to die, rain acid down from heaven. What need was there to turn Kalliste, 'the most beautiful,' into Thera, 'the isle of fear'? What need was there to destroy the Kallistens, that the smoke of their dying might persuade the pharaoh to let Moses and his people go?"

Past, present and future are only an illusion,
albeit a stubborn one.

—Albert Einstein

Time present and time past
Are both perhaps present in time future,
And time future contained in time past.

—T. S. Eliot, "Burnt Norton,"
in *Four Quartets*.

Oh, stop living in the past!

—Dad (to his paleontologist son)

BIBLE
STORIES
FOR
ARCHAEOLOGISTS

For a time, Thera stood at the crossroads of the Mediterranean world. Her ships transported goods back and forth to Cretan and Egyptian capitals in the south, to Attica in the north; westward to Sicily and southern Italy; east to Rhodes, Cyprus and Turkey.

The bull from the sea changed all that.

Among pottery fragments and stone building foundations lying on top of the Theran pumice deposits, we find unmistakable evidence that Minoan art and architecture continued to thrive, but in a hybrid style that reflects direction by Greek and Italian chieftains. Surviving scientists, architects and artisans must have set sail from Crete and the Cyclades, seeking new homes. If so, they sailed no longer as master traders of the Mediterranean. They sailed as subservients, members of a shaken and eclipsing civilization, with only themselves and their skills to offer. It is even possible that mainlanders arriving on Crete and sifting through the damage spirited the skilled and learned away to Italy and Greece, much as Germany's rocket scientists, valued by the Russian and the American victors of World War II, were snatched up and divided between them.

One remnant of Minoan civilization is believed to have sailed southwest to Tunisia, where they settled in the Atlas Mountains and became the tribe known to the classical Greeks as the Atlantes. Another remnant went south, where a handful of talented refugees apparently became Egyptian nobles; and others almost certainly sailed east, bringing their distinctive architecture with them to the land they called Philistia, or Palestine, in what is now Israel's Mediterranean coast west of the Dead Sea. Around this time, the Egyptian pharaoh Tuthmosis III was at war with Syria, and the first of at least six influxes of Jewish immigrants from Egypt to Palestine began. In Israel, intermittent layers of discarded Hebrew tablets and broken pottery run at least three hundred years deep, and are sometimes mixed in with the residuum of Philistine settlers. There were probably several exodes, several migrations of Jews into Israel—which explains, perhaps, why the Old Testament is notoriously unclear on the time of the Exodus, and does not associate the oppression with a specific pharaoh.

The late paleontologist and Jesuit priest Teilhard de Chardin (1881–1955) described those days as reaching a kind of critical mass, in terms of human population density in and around the Mediterranean.

"Sooner or later it was bound to happen," Teilhard said. "The very roundness of the earth, combined with gradually lengthening human lifespans, forced proximity and convergence on the human mass upon the planet. Originally, and for centuries, there was no serious obstacle to the human waves expanding over the surface of the globe; probably this is one of the reasons explaining the slowness of their social evolution. Then, from the Neolithic Age* onwards, and beginning on the eastern shores of the Mediterranean, these waves began to recoil upon themselves."

---

* The closing chapter of the Stone Age, characterized by the use of polished stone tools, the emergence of agriculture and the first use of pottery.

And recoil they did. According to Amos, a minor biblical scribe of the eighth century B.C., at least three waves of migrating people had converged on the same limited tract of land at the same time: Israelites (Jewish immigrants) came out of Egypt; the Philistines came from Caphtor (Crete); and the Syrians from war-ravaged Kir. The biblical record, which echoes back through documents dating at least to fifteenth-century B.C. Palestine (then frequently called Canaan, "the Promised Land"), provides glimpses of conflict and chaos almost from Day One—conflict that still persists nearly thirty-six hundred years later. Amos (5:7) has linked the migrations and confusions to a day when lands melted and the sea rose up to terrorize the earth:

> And the Lord God of hosts is he that toucheth the land, and it shall melt, and all that dwell therein shall mourn: and it shall rise up wholly like a flood. . . . It is he that calleth for the waters of the sea, and poureth them out upon the face of the earth. . . . Have not I brought up Israel out of the land of Egypt? And the Philistines from Caphtor, and the Syrians from Kir?

Geologists tell us that water truly did pour out upon the face of the earth (or at least upon what must have seemed to Mediterranean peoples to be all or most of the known earth). On the west coast of Turkey, just north of the island of Rhodes*, is a small body of water whose shoreline is like an ever-narrowing funnel. Its open mouth faces west, toward Thera, and anyone living behind that mouth might just as well have been a flea located in the throat of a cannon. As the shock wave surged east between increasingly confined shorelines, the waters piled higher and higher until at last they became a foaming white mountain eight hundred feet tall. The wave penetrated thirty miles inland, in the general direction of Mount Ararat; and when it receded, it dislodged house-sized boulders, scoured the soil and carved out chan-

---

* Where local lore has it that a city called Cyrbe was "swallowed up by a great flood-tide."

neled scablands. Elsewhere, on a strip of Turkish coast only ninety miles north of the funnel, the wave seems to have risen barely twenty feet high. Tsunamis are like that—capricious.

In ancient Greek literature there survives a tradition of devastating upsurges from the sea, usually attributed to punishment by Zeus or Poseidon for one offense or another. On the east coast of Greece, Athena and Hera challenged Poseidon for the cities of Attica and Argos. In his anger, Poseidon "submerged the cities under sea water." The ruins of Attica, much like the Turkish scablands, lie behind the Thera-facing mouth of a funnel. An epic poem of the period provides what looks for all the world like a scientifically accurate account of an approaching tsunami. Such waves simply rise up from the deep sea as they ride into coastal shallows, blotting out the horizon and showing no visible crest until they pile up so high that friction with the earth's surface causes the wave bases to slip and drag behind. The towering monsters then stumble forward, appearing to seethe and foam:

There came a sound, as if from within the Earth
Zeus' hollow thunder boomed, awful to hear.
The horses lifted heads towards the sky
And pricked their ears; while strange fear fell on us,
Whence came the voice. To the sea-beaten shore
We looked, and saw a monstrous wave that soared
Into the sky, so lofty that my eyes
Were robbed of seeing the Scironian cliffs.
It hid the isthmus and Asclepius' rock.
Then seething up and bubbling all about
With foaming flood and breath from the deep sea,
Shoreward it came to where the chariot stood.
—Euripides, *The Hippolytus*

Soil deposits from the Cyclades, Crete, Turkey and Egypt, and deep-sea drilling samples from points all around the Mediterranean, tell us that ash from the Thera eruption spread east, apparently following prevailing autumn winds. The work of the Greek dramatist Euripides, who lived on the west Aegean between 480 and 405 B.C., described a catastrophe that differs from east Aegean descriptions of prehistoric inundations in not associating the surge with melting lands, darkness at midday and deadly black clouds. According to the oral tradition handed to Euripides, a wall of water heaved up from the deep, into a sky as clear as glass. This is precisely what one would expect if the poet had recorded the memory of Bronze Age tsunamis in the western Aegean, which was spared the added calamity of Thera's death cloud. Euripides' mention of fearful thunder preceding (by some un-specified interval of time) a giant wave is especially compelling because the sequence of events mirrors what should have occurred, and the explosion of Thera would certainly have been audible in Greece. The blast wave from Krakatoa, in A.D. 1883, was heard more than two thousand miles away, and was powerful enough to shatter windows and crack walls at a distance of a hundred miles. The Greek coast was only two hundred miles from Thera, and Thera was at least six times as powerful as Krakatoa.

In the fifth century B.C., the Greek historian Diodorus sailed east to the Turkish island of Samos, where he reported that the inhabitants followed an ancient tradition of offering animal sac-rifices on altars that had been planted high above the beaches, to mark the flood line of a terrible inundation from the sea. According to legend, there had been a fire in the sky, and thunder and darkness, and Poseidon was said to have turned against his own children, who inhabited an island far away in the west; and on that island he covered up a city, "shutting it in with a mountain." West of Samos, the deserted streets beneath Thera tell the same story: a city truly was shut in with a mountain.

Within two or three generations, the memory of the buried city grew faint and hazy, then fainter and fainter, until even its name was forgotten. Only the island's ancient names, Strongyle and Kalliste, survived. Strongyle meant "the round one," though the island was no longer round. Kalliste meant "the most beautiful," though the vegetation that sprang up over the graves of Kallisten forests was ragged and thin.

The men and women who came to the island afterward built mudwalled houses and terraced farms on the dry volcanic soil. Their civilization was on the wane, and the volcano did not let the people forget who had put them on that course. It erupted and buried their farms, and others resettled the island—repeatedly. Each time, after a century or two of quiescence, the giant awoke to shatter their dreams. In A.D. 365, the northern coast of Crete was inundated by Theran tsunamis. They spread to Alexandria, Egypt, where ships rode on a surge of leaping, hissing water through the streets of the city. In A.D. 1650, a submarine explosion off the northeast flank of Thera caused the sea to withdraw suddenly from Crete, exposing wet mud that had, only minutes earlier, lain under twenty feet of water. At the keels of stranded ships, puddles glistened in the sun. Fish thrashed in them. The sea drained completely from Iráklion harbor, making sucking noises as it went. When it returned—all too swiftly—it was fifty feet above flood level. In 1672 Thera quaked and boomed again, hurling a wave over the top of Kos into the Turkish funnel. The shock of 1956 reached magnitude 7.8. This time, the damage to Crete was minor, but Amorgos and Astypalaea, fifty miles east of Thera, were pounded by waves up to 120 feet tall. Yet even here, the tsunamis were unpredictable. On parts of those same islands, they reached heights of barely six feet.

Through all those years, the civilizations that swirled around Thera were equally unpredictable and equally violent. In Greece, Italy and Egypt, massive and splendid architecture rose up on columns, then passed like a dream. The Goths and Vandals severed the aquaducts and never restored them. Colosseums and public baths fell away stone by stone, their features softened by wind and rain, while the inheritors of Greece and Rome drank from the same muddy rivers in which they washed their clothes and spilled their sewage. Christian tribes chiseled "pagan images" out of existence; almost all of the great frescoes and sculptures lying on the earth's surface were lost. A wave of Islam came out of the African deserts to the south and east of Thera, crossed west and converted the Atlantes tribe to the new faith before rushing across Gibraltar into Spain. There they built fair cities and enjoyed three hundred years of civilization, until a wave of Christian Crusaders struck out from the east, met the wave of Islam from the south, overpowered it and continued on toward Jerusalem.

The Mediterranean world could not hold all the warring factions. They flowed north and east, west and north, across lands whose forests had already been devastated by armies of shipbuilders. Thera, though still geographically at the center of Mediterranean history, was now a mere bystander. The island changed from Egyptian to Roman hands, to the Vandals and the Huns. It was subsequently owned by pirates, Byzantines, Venetians, Greeks and Turks. And yet while even the Pyramids crumbled into mere ghost images of their former glory, a city slept unsuspected within the earth, astonishingly intact, waiting only for the twentieth century and Spyridon Marinatos.

And the Earth was without form, and void, and
darkness was upon the face of the deep.

—Genesis 1:2

MEDITERRANEAN
GENESIS

WINTER, A.D. 1989

History has always surrounded Thera; and because the island is so firmly embedded in the ebb and flow of civilizations, it is, more than almost any place on earth, the perfect time probe. To most people two thousand years is an enormous amount of time. My training is in paleontology, and my favorite period is the age of the last dinosaurs. To most paleontologists, I work with the really young stuff—"Late Cretaceous? Paleocene? Cradle robber! Your fossils are so young they still smell!" they say. "Charlie goes back only two percent into the history of the earth."

To understand the story of Atlantis, you must understand the story of time, and man's little niche in it. You need perspective, and the only way I know of giving a true perspective is to begin slowly, a year at a time, and to move backward in increasing steps—first one year, then two years, then four—doubling each time. We will stand on Thera throughout history, stepping off from time to time to look at the Mediterranean world that encloses it, including Egypt, Rome, Greece, the Cyclades and Turkey. As we set out on our backward journey, an initially slow perspective will show us human change against a changeless earth. Slowly, the works of humanity will peel away; then, more and more rapidly, the positions of land masses will begin shifting, shifting like wax, and the very stars in the sky will begin to drift and then to race. As historical time converges with biologic time (the evolution of man), and biologic time speeds up to converge with geologic time, a fast perspective will show us Thera and a changing Mediterranean against which all human works evaporate out of existence.

We are about to embark on a roller-coaster ride, of sorts, a ride through the giddying depths of time itself. Along the way we are going to learn a great deal about where Thera and its inhabitants came from, but it's more than a mere tour of wonders. When you step off the roller coaster, you will discover that the ride has given you the most important archaeological tool of all. It is a tool of mind, a tool of imagination, a tool I like to call the *time gate*.

One cannot talk about Thera without dates rolling off the tongue: A.D. 1932 . . . A.D. 1902 . . . the fourth and sixteenth centuries B.C. . . . the millions of years it has taken mountains to rise up on the mid-Atlantic Ridge, and then to subside slowly under their own weight. But it is not sufficient merely to memorize dates, even if we think that we understand time. I want us all to really *feel* time. If our ride is successful, you will emerge from it as if into a whole new world. You'll have a new perspective on history that should stay with you not only through the rest of this book, but through the rest of your life. Thereafter, wherever you find yourself—on an ancient Theran street or in the shadow of Cheops—you will be able to reach out, touch a cornerstone, and take a grandstand view of time.

Thirty-six hundred years . . . forty-six hundred years . . . forty million years . . . as a paleontologist, I am used to traveling back sixty-five million years from midtown Manhattan to an upper Cretaceous sea (which reminds me, have you ever noticed that the walls of the Empire State Building lobby are polished slabs of rock cut from a fossil coral reef?) But most people find that the mind fails to deal with numbers beyond two thousand, fails really to embrace the heft of time in a rock. The year 1600 B.C. might just as well be forty million B.C. Without the time gate both dates daunt the imagination equally. Minoan Thera and fossil clams embedded in Cheops are merely very far away in time, much as Crete, the Pillars of Hercules, and the Atlantic Ocean must have seemed "very far away in the west," with little or no perspective to each other, as viewed from a vantage point in ancient Egypt.*

* Even at the Pyramids I was unable to resist the urge to travel back beyond civilization. That's how I darned near got shot by Prince Charles's bodyguards last summer.

I was searching for and photographing (and occasionally collecting) fossils in Cheops's limestone blocks. The Egyptian Army had cleared everyone out and sealed off the area prior to the prince's visit, but apparently I was walking around one side of the pyramid while they circled around the other. I did find it curious that I was the only soul in sight. About twenty minutes before Prince Charles arrived, I turned a corner and was immediately apprehended by men wearing berets and holding machine guns. One of them materialized from the stone blocks only twenty feet overhead. They realized very quickly that I was not a potential assassin. Still, they held me under guard. The prince arrived (with an entourage of TV camera crews, and twenty black Mercedeses full of men in long, black, traditional Arab robes), walked halfway to Cheops, posed there for about thirty seconds, returned to his limo and was driven off toward the Sphinx with an army escort leading the way, and more than a hundred dignitaries and cameramen trailing behind. The guards kept watch over me for about two hours, while army sirens whooped and screamed out in the desert, in the direction the caravan had gone. They were, all things considered, very friendly (even insisting that I take a picture), and they did seem truly fascinated to know that the Pyramids are marine limestone—though I am afraid I told them more than they ever wanted to hear about this subject.

They asked me to explain why I had behaved so strangely, why I had been walking around Cheops with my nose to the stones. So I showed them the shells. "You can spend days and days cataloging the different kinds of sea life in these blocks, trying to figure out what was going on." I held out a fossil clam, no more than an inch wide and looking no different from clams living in the Mediterranean today.

"The water here was clear. No land-washed river sediments were mixing in with it, just the skeletons of microscopic animals accumulating to form beds of the clean, white lime the Egyptians found so attractive for building pyramids. Forty million years ago all of this was a sea. The water must have been deep, but it was also sparkling clean. The sunlight got down to the bottom."

Hurtling backward through history, following the chain of cause and effect from its end to its beginning, may seem unnatural to most people; but to an archaeologist or a paleontologist it is the most natural thing in the world. As a general rule, the further down you dig, the further back you are looking in time.*

Tracing the record of man and earth down into time can be like leafing through the pages of a book, except that the nature of the pages is always changing. Our recent history is written on reels of film and sheets of newspaper. Go back only a few thousand years and you are rummaging through sheets of rock. When you come right down to it, almost everything we know about the planet is written in stone.

It is only fair to warn you, before you read any further, that this is a journey with no end, but it does have a specific beginning: 25 degrees, 25 minutes and 56.9 seconds east longitude, 36 degrees, 20 minutes and 24.2 seconds north latitude—which is where the buried city was located on New Year's Day, A.D. 1989. Thera is the vantage point from which we will look out upon a changing world; but latitude and longitude are only a temporary description. Like a fly embedded in amber, Thera is a prisoner of time and space. The island and the Middle Eastern, southern European and African shores that surround it will drift like icebergs on a sea, enveloped in the history of the planet. Stepping back by powers of two, we will follow a single spot on the earth down through the Minoan Empire, down through the Ice Ages and the dinosaurs, to the shard of time from which all things began.

"How can you know that?" a guard asked.

"Cheops is full of clams. That's what the clams tell me."

Cheops was built forty-six hundred years ago (give or take two hundred years). Most people think the Pyramids are old, but if you look at them closely you can see oceans going back millions of years.

* One exception to this rule is the rim of Meteor Crater in Arizona, where whole layers of rock were curled up and overturned by a piece of celestial buckshot. Another exception is San Marino, where the mashing together of continental plates has literally turned a country upside-down.

SPRING, A.D. 1988

There were five billion of us on earth. In that year, two dozen of us had lived outside the planet altogether, and dozens more had descended nearly three miles to the deep range of the ocean basins. If you wanted to, you could buy a ticket and fly to the antipode of the world in a single day. Humans moved freely around the world, and under it and out of it. None of these things were particularly newsworthy, unless you happened to look out the window of a transatlantic 747 to consider the long, hazardous voyages of discovery and colonization that had force-fed the emergence of a civilization capable of stripping aluminum from mountainsides and hurling jet aircraft into the sky. Those few who did take pause to look out the window were often asked by flight attendants to draw the shade, because the light was interfering with other passengers' efforts to take a nap, or watch the movie. The most amazing thing about living in those times was that so few people were amazed.

In April 1988, the TV cameras of the world focused on a new and, as far as the news was concerned, amazing aircraft. The seventy-pound, human-powered *Daedalus* flew seventy-four miles from Crete to the coast of Thera. From the air, a bloom of microscopic algae on the Aegean was horribly visible. It appeared to be expanding out of control, a hundred-mile stain on the earth, the color of drying blood. It was part of a global epidemic of algae blooms. Ted Smayda, a professor of oceanography at the University of Rhode Island, had emerged from a conference in New York and declared, "This is not a little blip that is going to go away. We are suddenly having an array of blooms of species which previously were unknown around the world."

The blooms were one of a growing number of warning signals, sometimes written as streamers of blood red water, other times as ozone holes opening in the sky. There were five billion people on the planet. In forty years there would be ten billion; but already the oceans and the forests had been wounded and were perhaps dying. The message from nature was clear: even five billion were beginning to push the limits.

WINTER, A.D. 1986

Halley's comet returned, as it had returned every seventy-six years throughout all recorded history. But this homecoming was different. If there had been sentient beings on the comet, they would have noticed unusual stirrings since its last approach in 1910. After 4.6 billion years of quiescence, the earth suddenly began emitting radio wavelengths—which grew brighter and brighter each passing year, until they outshone even the sun. Now, as the comet swung closer, titanium and silicon were molded into new shapes. The shapes were christened *Vega* and *Giotto*, and were flung away from earth. Directly they arrived and darted beneath Halley. One almost crashed into her. More pieces of the earth were ready to detach, more and more of them—space stations and shuttles and Mars probes—rising from the world like summoned spirits.

WINTER, A.D. 1982

On Thera, about a half mile south of the increasingly resort-oriented village of Fira, pumice miners came across upright cylinders of cold, black carbon poking through a layer of brown, loamy soil. The cylinders turned out to be the bases of trees, burned to cinders where they stood. The soil, which enclosed shadowy fossils of carbonized grass, had been the actual surface of the island three and a half thousand years ago. Handfuls of small, rounded stones suggested that a stream once passed through the site, and near the stream bed, the miners cut a cross-section through a fallen wall and fragments of crushed pottery. Further excavation was out of the question. A cliff of ash and rubble 250 feet tall was poised directly over the find. Already it was dropping streamers of fine, white dust. If even a stone were removed without first removing the overlaying mountain of volcanic debris, an excavator might easily become a permanent resident of ancient Thera.

In Greece, 150 miles to the northwest, a democratic government still in its birth stages ruled over Thera and Crete. But the Cyclades, which had for decades brought wealth from summer tourists, were becoming fouled with raw sewage. Pesticides were accumulating in the bodies of fish, and the waters off the coast of Sicily had been depopulated by giant, floating canneries. Only the Tunisians had restricted the size of their fishing boats, and forbidden fishing in spawning grounds. As a result, they were bringing back fish three times as large as those caught by the Sicilians, who were now beginning to sail across the Mediterranean in their factories, to ravage Tunisian waters. It threatened to become a shooting war.

Far across the Atlantic, Brazilian settlers slashed and burned an area of Amazon rain forest larger than Greece, Crete and the Cyclades islands combined, while carbon-burning power production assailed respiratory and immune systems, rained defoliating acid down upon the northern forests, produced more carbon dioxide than the diminishing trees and marine diatoms could absorb from the air, and began slowly to raise the temperature of the earth.

WINTER, A.D. 1974

The world growth in population began to ease for the first time in three centuries. There were 3.7 billion of us. Two years earlier, Eugene Cernan and Harrison Schmitt had become the last Apollo astronauts to walk on the moon. They landed in a valley adjacent to a basin dug out by an asteroid. The asteroid had actually penetrated the moon's skin, providing an exit for thick, black lava. Now, potassium-argon isotope dating of the rocks Schmitt and Cernan had picked up revealed that the overflow from the puncture must have spilled into the valley and cooled there 3.7 billion years ago. The valley itself was a six-mile-wide fissure opened up by the impact. Unlike any place on earth, it remained geologically quiescent, a familiar landscape dating back billions of years. It was assailed by neither wind nor rain; the most dramatic events were the occasional splashdowns of meteorites that punched out little craters and stirred the dust. About two hundred million years ago, a large rock had tumbled down into the valley and come apart as it ground to a halt. Since then, the dinosaurs have come and gone on the earth, but on the moon, Split Rock has remained undisturbed, has failed to move so much as an inch.

Not so the island of Thera. Between 1974 and 1989, seafloor spreading had caused it to shift more than a foot east on the terrestrial globe. The island of Kea, one hundred miles to the northwest, shifted with it, carrying along the newly deposed Greek dictator Georgios Papadopoulos, who had been exiled under the same law he had used to exile ten thousand real and imagined political enemies to smaller and bleaker island prisons.

This was the year Spyridon Marinatos died.

WINTER, A.D. 1958

Sputnik 2 had become an orbiting coffin for Laika the space dog. Unseen, the German shepherd began to putrefy eerily at zero gravity, staining the cabin walls with deposits of dark matter. The American satellite *Explorer 1* soon joined Laika, while former bridge designer Al Munier gathered a little group of men together in a brownstone building not far from the White House, to discuss a proposal he had in mind for a crewed lunar lander. The brownstone was what passed at that time for NASA headquarters, and the proposal was beyond the ridiculous. Not even a rat had orbited the earth as yet and been brought down alive, much less a man.

On Thera, an earthquake had toppled half the buildings in the cliff-side village of Fira. Most of the island's inhabitants were in the process of moving out. Thirty years later, only six thousand people would live there—less than half Thera's prequake population. Spyridon Marinatos was one of the few visitors to the island in that year. He'd become somewhat obsessed with the volcano. In January he climbed the twelve-hundred-foot-high limestone protrusion known as Mesa Vouno. On top were ancient Egyptian and Greek ruins, modified more recently by the Romans. The new eruptions had provided Marinatos with a yardstick through time. Since the beginning of written records, only the volcanic cones now breaking the surface in Thera Lagoon had come into existence. At that rate—not exactly greased lightning—Marinatos figured that it must have taken the five hundred or so layers of multicolored volcanic deposits visible in the shorn-off cliffs beneath Fira at least a quarter million years to accumulate. And the cliffs were only a small sample of the mountain that had once been. The volcano went down thousands of layers more into the sea. At the rate things happened, Thera must have started growing millions of years ago.

As he strolled through a Greco-Roman amphitheater atop Mesa Vouno, Marinatos noticed that the limestone beneath his feet was studded with shark teeth and snail shells almost identical to the teeth and shells of their modern descendants in the Aegean.

Turning a tooth between thumb and forefinger, he decided that perhaps the fossils weren't very ancient after all; but a thought stopped him. From on high, he could see where the vein of limestone went: right under the volcano. It had to be older than Thera, millions of years older—which meant that the rate at which snails and sharks changed must be somewhere between geologic and dead slow.

In the six thousand years of earth history that some biblical scholars allowed, as estimated from the (exaggerated) lifespans of Isaac, Noah and their forebears, you could barely build a decent civilization, or even a small mountain. In hundreds of millions of years, however, you could breed reptiles from fish, and mammals from reptiles, and give them big brains, too.

WINTER, A.D. 1926

We have stepped back to our last stop in the most incredible century *Homo sapiens* has ever known. It opened with the ascent of kitelike, motorized aircraft, saw at its midpoint the probing of the ionosphere with rockets, and closed with the bridging of interplanetary space.

There were 1.7 billion people on the planet in 1926. One of them, Adolf Hitler, had recently completed an eight-month prison term for disorderly conduct. He wrote a book, *Mein Kampf*, and was released early on good behavior.

Almost four million radios were in use. The Ford Motor Company's twelve millionth car rolled off the assembly lines, and the first regular air mail service had been underway for two years. Thera, meanwhile, was nearly four feet closer to New York than it would be in 1989. Nobody on earth was aware of this fact. The ocean basins were completely unexplored territory, and talk of continental drift could easily end careers in the natural sciences. In an address to the German Geological Association, a meteorologist named Alfred Wegener noted that the coasts of North America and Europe, South America and Africa, fit together like pieces of a jigsaw puzzle. He began to suspect that continents were wandering about like huge slabs of rock drifting on the surface of the earth, and they were moving at a rate of only inches per year. He reasoned that if one added the inches up over millions of years, then during the Age of Dinosaurs America and Europe must have been one continent, and Thera was New York's next-door neighbor. He was right, of course, but that did not help him.

Wegener had already become famous as a record-setting balloonist and an Arctic explorer, but when he suggested that the continents had moved, he was promptly labeled a "madman" and never again found funding for his expeditions. Another thirty years would pass before anyone listened. Too late: by then, in frustration and obscurity, Wegener had died.

SPRING, A.D. 1862

Mark Twain was twenty-seven years old; Charles Darwin, aged fifty-three, had recently published his *Origin of Species*; Thera was seven feet closer to New York than it would be in 1989; and the world population reached one billion.

Robot probes to Halley's comet and submarines capable of descending three miles to the bottom of the sea were beyond the fantasies of even the wildest dreamer. Indeed, large portions of the earth's surface actually remained unexplored. North and south of the Mediterranean, where latitude and longitude converged on specific points at the earth's poles, the Arctic and Antarctic were vast unknowns.

Shipping was then the only way around the world, and an ever-expanding traffic demanded quicker routes between such places as London and Bombay. A passageway between the Nile and the Red Sea would allow ships to sail directly through the Mediterranean into the Indian Ocean without having to journey down the coast of Africa. The Suez Canal, then under construction, could cut travel time to the Orient nearly in half. At around this time, engineers discovered that pumice dust mixed with powdered limestone in the proportion of three to one made an unusually durable cement that did not crumble in sea water. A mining operation was begun at Thera, which, along with Greece, had only recently gained independence from Turkish Muslims. On the island fragment known as Therasia, mining activity at the base of the pumice layer was often hampered by rows of stone blocks. Whole buildings appeared to have been buried under the deposit.

The Suez Canal Company regarded them as a nuisance and ordered them destroyed.

At that time, a Greek volcanologist named Christomanos happened to be visiting the island, attracted by reports of the sea turning milky white and great masses of rock rising in the center of Thera Lagoon. He became the first to write about a Minoan settlement, in what appears to be the first recorded instance of archaeology in collision with industry. Before Christomanos could begin to learn what was being exposed by the Suez operation, the artifacts were either smashed or quietly spirited away. He barely had time to spot a layer of vase fragments more than a foot thick. What appeared to be two vaulted tombs were found—and plundered. Bronze tools turned up . . . two gold rings . . . the carbonized base of a tree . . . and the skeleton of an elderly man of apparently medium build.

The rings and tools were sold. The skeleton, which, if preserved to this day, could have told us much about Minoan medical practices, diet, even the kind of work the man performed, was broken apart and thrown away. No other Theran skeleton has since been found.

WINTER, A.D. 1734

Thera was a Turkish port, located some eighteen feet west of its 1989 position. The island had once been a paradise; but the volcano made a desert out of it. Though grapes and tomatoes thrived in the mineral-rich ash, water had a tendency to drain right through it. Plants left on their own, though well nourished, usually thirsted to death. There were no olives or cypress on Thera. In fact, there were very few trees of any kind. Water was drawn from deep wells, and plants had to be watered individually by hand, each day. The fields were buttressed and terraced with stone walls that mimicked China's picturesque rice fields, rising tier upon tier in the southern highlands. Farming the Theran wastes was not easy, but for a century or two, the islanders managed to produce one of the Mediterranean's finest wines.* They persisted, rearranged the land to their purpose, and ultimately prospered . . . until the volcano awoke again to remind them that Thera belonged not to man, but to itself.

George Washington was two years old and Catherine the Great was four. They were two of 760 million. The British explorer Francis Drake held the official world record for southward expansion of the human frontier. In 1578 he sailed through the Strait of Magellan into a storm and was pushed south into a body of water. He named it the Drake Passage. This was in the vicinity of Cape Horn, about twenty-four hundred miles from the South Pole at latitude 56° S. This official record was not as significant as the British pretended. Cape Horn had been settled by nomadic tribes thousands of years before Drake got there.

WINTER, A.D. 1478

The latitude barrier was insignificant, for even longitude was now an unexplored frontier. Fourteen years remained before Columbus would discover Haiti and the Bahamas. He'd cross the Atlantic believing he was circling the globe on a newer, shorter route to India. He'd therefore call the people he found there Indians, and they'd be known by that erroneous name ever after.

Michelangelo was three years old in 1478. Copernicus was five. Leonardo da Vinci had just invented the parachute and, reading up on the story of Daedalus, had begun experimenting with hang gliders. Handguns were becoming all the rage in Europe, and Caxton was preparing Chaucer's *Canterbury Tales* for its third printing.

In the wake of the Crusades, Venice had emerged as the strong-

---

* Theran wines were selected by the Vatican to perform one of the infamous cleanings of Michelangelo's frescoes in the Sistine Chapel.

est Christian sea power in the Mediterranean. Venetian rule extended over Thera, Crete and parts of the Greek mainland; but Thera was essentially without civilization. The eruption of 1472 had submerged part of the island and stamped flat almost every home. Subsequent raids by pirates and Islamic conquerors had reduced the island's population to three hundred. In 1480, the Duke of Crete would give Thera to the daughter of the Duke of Naxos as a dowry. Fifty-seven years later, the pirate Barbarossa, siding with the Turkish sultan, would sieze Thera and the Cyclades and cede them to the Turks.

The Venetian Empire was determined to rid the seas of both pirates and Islam. Venice had too much to lose. It was ideally located between the increasingly wealthy countries of western Europe and the Near East, sources of spices more valued than gold in Spain, France and England. To sustain her trade routes required ships, lots of ships, because the hostilities between Christians and Muslims had not ended with the Crusades, and showed no signs of ever ending. The Holy War was spilling over into the sea. All along the Mediterranean, whole forests were being felled to build navies.

In that year, western Europe and Venice were the technological masters of the world. But there were five hundred million people on the planet, only sixty-five million of whom lived in Europe. There were the Aztecs, the Incas, Polynesians and Japanese—a world full of disparate civilizations that had yet to find one another. Nobody knew, at that time, that two "superpowers" had emerged.

SUMMER, A.D. 966

The focus of human civilization had shifted. In the tenth century, China was the technological master of three hundred million people, having developed paper, gunpowder, the widespread use of steel and the first magnetic compass, yet it was Arab ships that ruled the Mediterranean Sea and the Indian Ocean. In Baghdad, Samarkand and Cordoba, Islamic scholars studied botany, medicine and mathematics. They preserved whatever Roman technology had been handed down to them, calculated the circumference of the earth and determined the moon's influence over tides. Tales of far-flung lands became a folklore that still exists today in the stories of *Sinbad the Sailor* and *Antar of Arabia*, the Bedouin Lancelot who braved wild beasts for a fair maiden's hand. Such epics, carried north by Christian Crusaders, would pave the way for medieval Europe's age of chivalry.

While the Arab empire spread Islam east to the people of Indonesia and the Philippines, Islamic scholars in Spain hand-copied what few books had survived the burning of the library at Alexandria by Archbishop Cyril in A.D. 415.* In these books they rediscovered Euclidian geometry, instructions for building mechanical gear shifts, steam engines, water-powered devices, and all the principles of Greek and Roman architecture.

At Cordoba the Arabs scavenged Roman ruins, demolished a Christian basilica and cannibalized the columns and masonry to build the mosque of a thousand pillars, each connected by arches and the whole supporting a vast roof. They installed street lighting, founded a university and built three hundred public baths. The rest of the Europeans, including the Therans and the Cretans, were in the midst of a Dark Age—troglodytes wandering through a post-Roman wasteland, defecating in public and wondering what the half-buried lavatories they found could ever have been used for.

Nothing of consequence was being invented. The time was unique in that everything the people knew or used was old. If the inhabitants of Thera had taken the time to climb Mesa Vouno and look closely at the amphitheater and the eroding pillars, they'd

---

* Under Cyril's direction the last scholars of Egypt's Alexandria (then under rule of the fading Roman Empire) were flayed alive to the bone with oyster shells. Their remains and their works were then publicly burned. Archbishop Cyril died of natural causes at an advanced old age, and was made a saint.

have wished they hadn't. They'd have understood immediately, without a word being spoken, that no one in those days could build anything like that.

It was a dull, depressing time for rational beings. Attitudes toward nature were still based upon superstition. Cats were said to be the embodiment of satanism and witchcraft, and thus began the Christian slaughter of cats (and "witches," too, who could be easily identified by cats' friendliness toward them). Such practices would inevitably make it easy for black rats to proliferate through Europe, carrying with them the Black Death.

Attitudes toward Muslims had been bad from the start. Though Muhammad had begun as a Christian prophet, even bowing toward Jerusalem each morning, Arab Christians tried to kill him. He fled, survived, founded a new religion and amassed an army. After his death in A.D. 632, the army moved out of the desert, swept west across North Africa, converted the Berber and Atlantes tribes of the Atlas Mountains to the new faith, and sent Christians fleeing across the Mediterranean to Sicily and Venice.

In A.D. 711, an army of converted Berbers crossed the Pillars of Hercules. They renamed one of its peaks Gibraltar, then moved on to found Cordoba, with its artificial lakes, fountains, parks, libraries and more than a hundred thousand houses. North and east of Spain, books and architecture were almost unheard of. Amid Roman roads that led nowhere, and amphitheaters with no performers, people had their work cut out for them just keeping fed and warm. They were an uncouth, unshaven, smelly lot, prone toward boozing, brawling, killing witches and pillaging (and that was just the women).

And then the pope announced that all true Christians should take up arms on a Crusade "to rescue southern Spain and Jerusalem from the clutches of the evil, perverted, treacherous, sadistic Arabs who have held those regions for hundreds of years." That was the official Christian version of what the Crusades were all about, as put forth by a pope who prayed to his dear God that the mobs would be just as happy raping and pillaging somewhere else. To make the endeavor even more attractive, he borrowed something Muhammad had once said about a jihad, or holy war: that true believers who were killed while carrying the word of God would be taken directly to paradise.

Spain fell. Then the Atlas Mountains. Then Libya. Finally, in A.D. 1099, a Crusader army—a rough beast of an army whose hour had come around at last—reached Bayt Lahm, on the outskirts of Jerusalem. The Crusaders fought among themselves over which group of soldiers should take charge over pillaging the city when it fell, and then swiftly moved in. They celebrated with one of the most brutal massacres history had ever seen. Muslims and Jews—wounded soldiers, young scholars, women with infants crying in their arms—were rounded up, shepherded into mosques and synagogues, barricaded in and burned alive by the thousands.

The savage brain had forgotten—if indeed it had ever learned in the first place—the lessons of tolerance and mutual tenderness set down by Christianity's founding prophet, who had called peacemakers the children of God. On the night the Crusaders entered Jerusalem, an orange glow went up into the sky, and as a hot wind roared up from the south and the flames burned brighter in its fury and voices cried out and were cut short, the conquerers rejoiced that Christianity had at last returned to the Holy City.

WINTER, 58 B.C.

The distance between Thera and New York was shorter by 110 feet. The world population was roughly the same as it would be a thousand years hence—three hundred million people. There were no holy wars between Muslims and Christians, for neither religion existed as yet.

The Ptolemies of Egypt (the family to which Cleopatra belonged) had made Thera a major naval and military outpost. They built a fortified city atop Mesa Vouno, the mountain on which

Spyridon Marinatos would stand two thousand years later, contemplating deep time. If you stood on that mountaintop in 58 B.C. and looked very carefully at the night sky, you'd notice a slight change. Alkaid, the star at the tip of the Big Dipper's handle, was 184 seconds of arc southeast of its 1989 position. You could almost see the difference with your naked eye. It was equivalent to a pinhead held out at arm's length.

The year 58 B.C. was the time of Rome's ill-fated triple pact, formed by Pompey, Crassus and an ambitious little fellow who in his youth had been forced to seek a new hiding place almost every night, and to bribe householders to protect him from dictator Sulla's secret police. Pressured by public opinion, Sulla relented with these words: "Very well then, you win! Take him! But never forget that the man whom you want me to spare will one day prove the ruin of the party which you and I have so long defended. There are many Mariuses in this Julius Caesar."

The world had two technological masters, Rome and China, each so far from the other as to preclude dreams of conquest, or even viable trade. The Roman superpower encompassed almost seventy million people, only slightly fewer than China, although industrially the Chinese appear to have outclassed Rome.

Rome owed all of its technology—the water wheels that ran its mills, its knowledge of metal, the printing press in Herculaneum, the steam engine Nero would mount on his barge—to the Greeks, whom they had succeeded as masters of the Mediterranean, and who in turn owed much of their technology to the preceding Minoan civilization. During the brief interval in which they held sway over the western world, the Romans invented very little that was new; but it was their ships, it seems, that were the first to land in America.

In A.D. 1982, scuba divers located a shipwreck dating to the first century B.C. in Rio de Janeiro, Brazil. The wreckage contains several hundred long-necked urns with distinctive handles. The urns were used to carry water, wine, oil and grain on long voyages. They are called amphora and—there can be no doubt about this—

they were manufactured in Rome. The ship was found in a sheltered bay, which suggests that it was maneuvered there by a crew that had survived the journey.

In 1987 a second shipwreck was discovered in Venezuela. Again, its cargo is undeniably Roman. Given the exceedingly low probability of any single shipwreck ever being unearthed, the Brazil and Venezuela finds must represent only a small fraction of the vessels that actually made it to America (surely there will be other finds). Given an empire that endured for centuries, one need only have a single ship stray off course every couple of years to have hundreds of them America-bound. *

The next obvious question: if dozens or hundreds of Roman ships landed one by one in America, did any of them ever return to tell the story? If news of a continent opposite the Pillars of Hercules ever was returned by seafaring Romans, we will never know. Written records, if such existed, were lost with the library at Alexandria.

We can easily imagine such news being immediately associated with the legend of the lost continent. Plato had, after all, written that the bones of the land still remained. It is even possible that some of the ships that ended up in South America had in fact deliberately set out to search for remnants of Atlantis—which puts Plato in the curious position of having guaranteed, sooner or later, the discovery of America.

WINTER, 2106 B.C.

The buried city of Thera was now in daylight; and the Minoan Empire had begun to establish island colonies throughout the eastern Mediterranean. Its ships traveled to Egypt, bringing wine, olive oil and cypress wood for trade. At Alexandria ideas, too,

---

* Only a hundred miles west of Gibraltar, ships entering the North Equatorial Current will be aimed, like Columbus, toward Puerto Rico, the Antilles and Venezuela.

were being exchanged. At Knossos and Thera, multistoried buildings went up, without battlements or defenses. These were economic centers with courtyards and gardens, storefronts and bull arenas. Many of the roofs were edged with the two-pronged symbol of bull horns. Jewelry, too, depicted the bull. Bull images appeared repeatedly on Minoan pottery, and on frescoes. As a symbol of power, the bull was worshiped above all others, and vestiges of Bronze Age bull rituals would persist in Europe even forty-one centuries later, in Italian jewelry's lucky charm—a golden bull's horn—and in Spanish bullfights.

Thera itself was still whole, a circular island with a thickly vegetated volcanic peak that rose perhaps a mile into the sky. Even though Thera was only seventy-five miles from Crete, in centuries to come its art and architecture, though similar to works being created in the Minoan capital, would develop their own unique character. As part of a vast island culture, even if its isolation was incomplete, Thera was bound to breed diversity.* Across many different islands that competed with each other in trade, it must have been exceedingly difficult for Knossos to enforce social and intellectual conformity. But so long as the islands prospered, perhaps Knossos saw no need to bother. And prosper they did. Political power was essentially in the hands of the merchants, who no doubt promoted the innovations on which their prosperity depended.

The history of human progress has been a mixture of competitive and cooperative behavior: a dual need of equal and opposite strategies. Cooperation allows us to more effectively exploit na-

---

* It appears that when you rob people of the everyday world, you sometimes give them more than they might ever have had otherwise. From isolation Stephen Hawking sent his mind where no one had gone before—into the horizons of black holes. Arthur C. Clarke notes that he began sending his mind deep into unexplored territory during a two-year convalescence. This seems to be something the hermit philosophers of biblical times knew all along. New innovations may not always require a period of isolation, but it certainly helps. Even nature seems to work that way. Speciation happens fastest in small, isolated populations, especially in places like the Aleutian and Galapagos islands. The history of Thera and its surrounding islands suggests that similar rules may apply to civilization.

ture. Try to imagine one person taking on a mammoth. Hunting works better with bands. Agriculture works better with tribes. High technology works better with cities and nations, and technological growth is quickened when nations compete with each other on economic fronts, or even military ones.

Wherever universal empires have formed, seizing control of all people within reach of their horses and ships, stagnation and even regression have followed. The story of Egypt is one of impressive advancement over the few centuries leading up to Minoan times, followed by almost two thousand years of decline.

By 46 B.C., when Julius Caesar met Queen Cleopatra at Alexandria, Egypt had been in cultural eclipse for centuries. Art styles had gone essentially unchanged since the pyramid-building first dynasties, and the country had become impoverished. Julius Caesar's Roman Empire was about to engulf every nation within range of its influence, whereupon the Romans would repeat Egypt's history—centuries of stasis followed ultimately by decline.

After the long night of barbarianism ended, the focus of empire building would shift away from the Mediterranean and to the north. There, Napoleon, Kaiser Wilhelm and other would-be empire builders were going to find themselves frustrated to varying degrees by independent nations that knew how to exploit natural barriers such as peninsulas, mountain ranges, bogs, endless tracts of snow and all the other conditions that splitting and mashing continental plates are capable of producing—including, of course, the moat surrounding Britain.*

In the Minoan world, every island had the Aegean for a moat. Competition was inevitable, universal control virtually impossible. There was no reason for stagnation ever to set in. If Thera

* Robert G. Wesson, professor of political science at the University of California in Santa Barbara, points out that in eastern Europe—where natural geographic divisions are generally less clear-cut than in the west—nation-states have been less stable against takeover, there has been less political freedom, scientific and technological progress has been relatively slow, and multinational empires have prevailed.

had not exploded, there might have been television by the time of Christ.

Even after the Minoans disappeared, their island habitats fostered the diversification and innovation known as Ionian Greece. But transmission of this great revolution in human thought outside the islands was suppressed by the surrounding Greek Empire; Plato himself was one of the suppressors. In an affront to Zeus and all the other gods, an Ionian scientist named Democritus had dared to deny the existence of immortal spirits: "Nothing exists, except atoms and empty space," he said, and then went on to suggest that the Milky Way was composed of many stars, most of them too far away and thus too small to be seen by unaided eyes. Plato urged that Democritus' books be burned. No sooner was this done than another Ionian upstart named Aristarchus concluded from the size of the earth's shadow on the moon during lunar eclipses that the sun had to be much larger than the earth and very far away, that the earth was round and orbited the larger body and, as Democritus had suggested, that the stars were distant suns and there were an infinite number of worlds out there. Aristarchus' books, too, were recommended for kindling.

There is no evidence of suppression among the Minoans of 2106 B.C. They were surrounded by no one except themselves. When their ships sailed into Egypt, the Pyramids had, for five hundred years, been the wonder of the world. Cheops was still cased in a glass-smooth surface of polished white limestone. Its peak was gleaming gold plate inlaid with silver hieroglyphs wider than a temple door.

Four thousand years hence, some would claim that people from Atlantis had taught both the Egyptians and the people of Central America how to build pyramids, or that the Egyptians had crossed the Atlantic and taught the Maya how to build them. Though the existence of two Roman shipwrecks in South America indicates that such crossings were indeed possible, the Egyptians had stopped building pyramids by 2106 B.C., and had for centuries been burying their pharaohs in mountainsides. There is every indication that they had, by this time, forgotten the art of pyramid building. Twenty-five hundred years hence, Cheops's gold would long since have been removed, melted and sold off by scavengers. Even the jacket of smooth limestone would be gone—stripped off piece by piece to build crude dwellings, or to be burned as lime. Egypt itself would consist largely of disorganized nomadic people, unaware of their former glory. That's when the great pyramids of Mexico would start rising out of the jungle.

WINTER, 6202 B.C.

We have crossed into prehistory. No battles, no kings, no explorers, no centers of population have left names behind them.

If you looked up at the night sky, you'd notice that the handle of the Big Dipper appeared to be straightening out. The climate was colder than it is today. The latest Ice Age was ending, and there were probably fewer than thirty million people in the world. No one knows if any of them lived on Thera. No tools or pottery dating earlier than 3500 B.C. have been found on the island. If they exist, they will be difficult to locate, buried as they are under thick layers of Bronze Age ash.

Life was intensely local. Most people lived and died in the same mudwalled hut, never venturing farther afield than seven miles (which happened to be only eighty-four times farther than Thera would move on the map between 6202 B.C. and A.D. 1989). In Europe, Africa and Asia, populations tended to be so isolated from each other that even when new ideas did originate, there was no way of transmitting them.

The Polynesians were the technological masters of the world. For two thousand years they had been spreading across the Pacific in canoes, navigating by stable coordinates in the sky—stars. Radiating out of Indonesia and Australia, they had extended their range to New Zealand, where they began to hunt the giant eagle and the flightless, twelve-foot-tall Moa birds to extinction.

People in Thailand were growing beans and manufacturing wood-fired pottery. A pottery industry had also developed in Japan. Cereal grain was being grown in Iran, and in central Turkey, about five hundred miles northeast of Thera, the city of Catal Huyuk (a name given to it by modern archaeologists) had emerged. It was among the first settlements of its kind and it is there, or close to it, that we can probably trace the origin of Minoan bull-worship, if not the bull-worshiping Minoans themselves. Six thousand people lived at Catal Huyuk within a span of thirty acres. Mud-brick houses were packed shoulder to shoulder, leaning against one other for support. A third of all the rooms excavated at the site appear to have been shrines dedicated to the worship of bulls. Everywhere are altar-like constructions, clay images of bulls, bull horns built into walls, and occasionally offerings of human skulls placed beneath carved bull heads.

Being the most powerful animal around, the bull apparently became not merely an object of the hunt, but an object of worship as well—the sacred incarnation of power and strength.

As these people built shrines, scattered seeds on arid soil and set off on hunts in the mountains of central Turkey, the bones and eggshell fragments of animals mightier than bulls were eroding out of cliffs and getting trampled underfoot. Later generations would call them giants' bones. Augustus Caesar (Roman emperor, 31 B.C. to A.D. 14) was going to have whole skeletons excavated from the rocks. He'd build the world's first dinosaur museum at Caprae, and the Bible would note that "there were once giants in the earth." Few took note of the impressions of broad-leaved evergreens in the rocks that enclosed the bones—hints of a green world alive with bellowing, honking and chirping giants, a world that had vanished millions of years ago.

WINTER, 14,394 B.C.

Taurus Littrow Valley looked exactly as it would in A.D. 1972, before Eugene Cernan arrived and sketched his wife's initials there. On the earth below there was ice. It piled up more than two miles high over northern Europe. England and Germany were sinking under the weight of ice sheets. The earth's crust dunked down like a ship loaded with heavy cargo. Scotland had dropped into the mantle. Its mud was fusing into shale. In eight thousand years, when the cargo of ice melted off, Scotland would float up again, raising beaches out of the earth.

Some of the Aegean islands, but apparently not Crete and Thera, became hills on uninterrupted European plains where now there is water. The water that formed the giant ice caps came from the sea, and their formation had lowered global sea levels by almost three hundred feet, causing new lands to rise out of the Aegean.

Thera itself remained an island paradise. The last major eruption had occurred about 23,000 B.C., adding at least four new cones to what was already a complex central peak. The mountain had remained quiescent ever since, and its forests supported a strange assortment of animals, including elephants as small as dogs, and dormice as large as cats.

The world human population was probably below eight million. No one had yet reached Thera or Crete, because pygmy elephants continued to survive there, and the fossil record suggests that wherever humans went, elephants of all species declined quickly toward extinction.

WINTER, 30,778 B.C.

Alkaid, the star at the tip of the Big Dipper's handle, was 2,949 seconds of arc southeast of its A.D. 1989 position. You would easily see the difference. Its displacement was equivalent to the diameter of the moon.

Southern Europe was mostly a rolling tundra of knee-deep grass and waist-high sedge. Near Cordoba, Spain, and in southern France (where Ice Age people were painting bulls in caves*), on the Mediterranean islands and across Africa, the forests continued to thrive. The Theran forests, large portions of them, were being burned and flattened under flows of lava. In the center of the island, a new period of mountain building was underway. The climate in Upper Egypt was less arid than today's. Hippopotamus bones found lying under desert sand three miles away from the Nile are evidence of swampy areas where none presently exist.

From what we know of the earth's ice ages, its fossil records, subtle markings on the Martian ice caps and varying levels of sun-dependent carbon isotopes in the rings of the oldest trees, our sun appears to have been acting oddly over the past seventy million years. The dinosaurs, and most of the insects that lived with them,† were the first casualties of change. Life seems to have become even more challenging during the recent Ice Ages, but in those days there were not five billion, and more, well-armed human beings squeezed border against border from pole to pole. If ice and drought did threaten a population, there were unin- habited places to move to. There were new resources for the taking, virgin continents to be explored; plenty of baskets for humanity to put its eggs in.

And so, it is with a sense of alarm that astronomers of the twentieth century have turned their attention to the sun, Alkaid, and all the other stars in the sky. A desire to learn more about

---

* There is no evidence that people were actually living in the caves. The sites were apparently gathering places for bull rituals—which included painting. Cave men are just another popular story that never happened. By 30,778 B.C., people had already been building huts for two hundred thousand years. Caves are perpetually moist and uncomfortable, and people simply prefer not to live in them.

† A census of fossil insects I conducted in 1978 suggests that this class of animals, long believed to have come through the dinosaur extinction unscathed, in fact suffered as severely as the vertebrate classes.

the activity cycles of stars is no longer rooted in mere childlike curiosity. The past teaches us that ice ages have come and gone, and will surely come again.

SPRING, 63,546 B.C.

Earth was still in the grip of the Ice Ages; but 63,546 B.C. marked a few moments of respite. The world was entering a warm interglacial period. The ice retreated. The rivers discharged as much as four thousand cubic miles of meltwater in a single year, raising the oceans two inches. On the steeply sloping shores of Thera, the rise was barely noticeable; but on the broad, flat continental shelves, the sea lapped menacingly at coastlines, advancing inland two hundred feet per year.

This was the era of Neanderthal man. He is traditionally depicted as the original club-bearing "ape man," a brute who dragged his women around by the hair, lived in caves and had hideous table manners.

Iran's Shanidar burials, dating to about 58,000 B.C., provide a more civilized picture of the Neanderthals. One individual had lost an arm and the use of one leg to arthritis. He was unable to take care of himself, yet somebody sheltered him and fed him for several years before he died. And when he died, somebody grieved for him and placed him gently into a grave and provided him, even in death, with food and tools.

At a number of burial sites in Europe and the Near East, bodies were placed on their sides, with the legs drawn up, and with the head resting comfortably on the right arm as if death were regarded as a kind of sleep. The graves contain finely crafted flint tools, charred animal bones (indicating roasted meat) and fossilized blobs of pollen (the remains of flowers). It is not likely that the dead would have been provided with fine tools and cooked food unless these things were somehow meant to be used.

The story the bones tell is one of deeply felt loss and a belief in the persistence of consciousness, or the human spirit, beyond death. The Neanderthals were probably no less human than any of us living today—brutes that never were.

Their bones, however, differ from ours in being relatively heavy, albeit fully human in structure, and their muscles must have been correspondingly heavier. Human fossils with a pattern of Neanderthal features extend from Portugal across the mid-East and into Asia. Their greatest concentration appears to have been around the Mediterranean and in western Europe. There is no evidence that Neanderthals, or any other group of people, had reached Thera or Crete by this time.

The Neanderthal tribes appear to have been displaced by tribes of less robustly boned humans, sometimes called "archaic" *Homo sapiens*. These "archaic" humans might or might not have been the direct ancestors of "modern" *Homo sapiens*. In some scientific circles, modern man is believed to have diverged from an even more ancient stock, with *Homo sapiens neanderthalensis* occurring as a "side branch." We simply do not know yet. The evidence is too sparse, and often downright confusing. In Israel's Skhul cave, a skull dating from about 35,000 B.C. has a Neanderthal face tucked into an archaic human cranial vault. Skeletons from a neighboring site in Tabun predate the Skhul specimen by about five thousand years. They are "classic" Neanderthal. Whether the Skhul specimen represents an intermediate stage between archaic and modern humans, the product of interbreeding populations of both, or a side branch is anybody's guess.

What we do know is that Neanderthals became established in Europe and the Near East about 130,000 B.C. We know that they spread to the far eastern shores of the Pacific, and that they vanished abruptly around 37,000 B.C.

A modern-appearing skull from Borneo may date back as far as 38,000 B.C. At about that time, the first humans crossed the Exmouth Trench—probably in simple boats—to Australia. The oldest Australian skulls possess heavy brow ridges and projecting jaws displaying the characteristic Neanderthal gap between the wisdom teeth and the lower jaw's ascending branch. The full, "classic" Neanderthal face is not present, however, and the younger skulls that succeed these earliest-known Australians are modern.

It seems unlikely that the "Australian Neanderthals" evolved directly into more modern-appearing humans because migrants probably continued to filter in from mainland Java and Borneo throughout the last glacial period. There is no cause for believing that, once the first people arrived in Australia, they simply stopped arriving.

It follows that the story of Australian bones tells us much about what was happening on the shores of the Mediterranean. The Australian record is one of a constant influx of people who were undergoing a loss of Neanderthal robusticity in one or more places beyond Java. This record perhaps reflects an episode of outward migration (with Australia on the receiving end) wherein one breed—a newer, less robust one—somehow outperformed the others and won for itself an entire planet.

SUMMER, 129,082 B.C.

At Thera's shore, the roots of the mangrove provided a resting place for all manner of debris; decaying vegetation, driftwood, clam shells and stream-washed sediments. Beneath them grew an ever thickening mire—a darkening swamp of their own creation—until suddenly there was land where once there was water. During the three hundred centuries that preceded the rise of Theran mangroves, the whole surface of the earth had shifted and flowed. On Thera, and especially on Crete, forests were pressing forward against the rising waters of the latest glacial recession, advancing fifteen miles, and more, into the sea. Places where mangrove trees once stood literally in salt water would ultimately lie far behind the crests of what, in geologic time, was becoming a vaporous wave of outracing green.

An unusually warm interglacial period had begun. Along what would eventually be named the river Thames, near the place where London stands today, lions and hippos roamed, where "only" three thousand years earlier seals had pursued fish under pack ice. In terms of geologic time, the west Antarctic ice sheet fell like an express elevator, releasing imprisoned water and raising beaches all around the world. On Thera, clam beds flourished sixty feet above the present sea level, on the very spot where the buried city would one day be built. The island itself was 1.3 miles west of its A.D. 1989 position. The earth on which it rested, and the sun, and its neighboring stars, had swept in a great arc around the galactic center, covering a distance equal to 1220 years of light travel time.

Some of the earliest Neanderthal remains are tentatively dated to 130,000 B.C. These are scraps of skull representing two individuals from the Kibish formation, in the Omo, on Ethiopia's side of Lake Turkana. Both were recovered from the same horizon (that is, the same time frame) in the sediments; yet there are important differences between them. One is heavy of bone with the relatively low forehead characteristic of Neanderthals. The other is more lightly boned, with a more modern-appearing face.

The skulls appear to represent differences between two separate but contemporary populations. They form part of a picture in which new variants branched off from existing ones in odd, almost random directions.

If we pursue the vulgar version of Darwinism (not Darwin's version), then small changes in the shape of a skull must be explained as the perfect result of a long process of testing and refinement. This view forces some theorists to explain Neanderthal's large nasal cavities in terms of evolutionary usefulness. And explain they do: large nasal cavities were useful because they provided additional volume for warming air in a chilly climate. This might be true, but the same enlarged cavities, which gave the Neanderthal nose a broad, flattened appearance, are found in

Europe before the end of the warm interglacial period, and in equatorial Africa afterward.

Although a strict adaptationist might argue that nasal morphology could have become fixed during a previous ice advance, the counterargument is that people had been wearing clothes for thousands of centuries. An animal skin scarf could have done the job just as well and been genetically less complex. Enlarged nasal cavities ("useful" or not) might simply be a side effect of overall Neanderthal robusticity.

Peering back into the cellars of time—to European Neanderthals and the Omo fossils—we are reminded, perhaps, that ours is not a straight and narrow, connect-the-dots ancestry. Instead, we appear to be located at the tip of a branching pattern whose most outward sign is our present range of skin colors (with apparent adaptive advantages to specific climates) and skull shapes (harmless, random variation). Look back far enough and we may encounter any number of side branches converging on a single point at the base of the *Homo* limb.

WINTER, 260,154 B.C.

If you stood in the Taurus-Littrow Valley that year and compared its rocks and massifs against photographs taken in 1972, you would have seen immediately how changeless the moon can be. Somewhere in the valley, one or two or perhaps as many as a hundred fist-sized splash marks were missing. You'd have to look awfully hard to figure out which ones. Only the sky would tell you that anything had changed. To a visitor from futurity, the desert regions and ice caps on earth wouldn't look right. You could see this from Taurus-Littrow even without a telescope. And the stars? The Big Dipper was stretched out almost beyond recognition. It resembled a large chevron or spade, covering twice as much sky as it would in A.D. 1989.

SPRING, 522,298 B.C.

On the continental shelf west of Australia, a flying mountain had fallen down from space. A billion trees vanished in a searing white glare, and from the smoke of their burning—preceding the very sound of the explosion—came tight swarms of molten glass that had only an instant earlier been sand and stone. Today, black pearls of glass—pieces of the Australian countryside—can be found splashed across southeast Asia and Russia. The explosion put the Minoan eruption of Thera to shame; the sound of it alone must have been capable of causing permanent damage to one's hearing as far away as Thera.

An almost identical event is widely believed to have ushered in the end of the dinosaurs, yet no mass extinctions of life on earth are associated with the Australian asteroid impact. Everything seems to have gone on quite normally along the shores of the Nile, where the people we know by the name *Homo erectus* were carrying out organized hunts and occasionally dropping and losing finely crafted stone hand axes, in places that their distant descendants would come to call Cheops and Karnak.

These people had a skeletal anatomy similar to ours, except in the weight of their skulls, shoulder blades and limb bones. They were, in fact, more robust than the Neanderthals. There eventually arose three major tribes of human-appearing creatures on earth (*Homo erectus*, "Archaic" *Homo sapiens* and *Homo sapiens neanderthalensis*), each more different from the others than modern "races" or tribes appear to each other today. During the brief interval in which they lived as contemporaries, the combined population of all three tribes probably numbered fewer than three hundred thousand; so isolation and variation could occur, in the geologic sense, very quickly.

We know from diggings north of Cordoba, Spain, that bands of twenty to fifty individuals came together, set brush fires in strategic locations and drove migrating herds of deer, horses, wild cattle and elephants into the swamps. One fossil swamp bed contains hundreds of bones belonging to an extinct variety of straight-tusked elephant that stood higher at the shoulder than today's African species. It's all there in the sediments: tools of stone and bone, even wood, and widely scattered traces of burned trees and grass. Elephants' legs were severed and arranged in rows. No one knows why. One elephant had the entire top cut away from its skull. The brains were scraped out with stone tools and (probably) eaten. Bones were cracked open to expose the soft marrow. These, too, were scraped clean.

In Terra Amata, France, people were leaving their footprints on a beach that no longer exists. They trimmed saplings and leaves and fashioned from them temporary dome-shaped shelters. They built cooking hearths, they slept on animal skins and wrapped their bodies in them.

You don't do such things—build huts, control fire, organize bands of hunters to drive herds of wild animals into just the right place—unless you have a means of communicating very specific ideas. It begins to look as if our ancestors knew language a half million years ago. We can probably push its origin back farther still.

Minds that were then only beginning to awaken on the earth would one day speak of wood and gold and all other matter as frozen energy. And they'd explain that this is the reason uranium and other very heavy, unstable metals seemed to be leaking energy; and they'd scoop plutonium out of the earth, and speak of casing it in plastic explosives, and gold, and sodium, to create the unspeakable.

For millions of years, pressure very deep within the mantle had sent liquid magma probing up toward the Mediterranean at very high temperatures. All over Thera there were young volcanic rocks and hot springs. In the fluid, mantle rock, gold combined chemically with other elements, such as chlorine. Hot water circulating down into magma chambers dissolved gold chloride and brought it up. The heated solution sought out weaknesses in crustal rock and surged into cracks where faces met. There, the cooling gold divorced chlorine and precipitated out of the water as glittering snowflakes. Silver fell out of the water, too, and tin, and traces of uranium.

Thera's springs would one day be of great consequence to the descendants of elephant hunters along the edges of the Mediterranean. Long after the steaming fissures cooled, the Minoans would mine the metallic snow from them. When finally they abandoned the island, they'd take all the gold and silver with them to Crete. From there, Theran gold would disperse anonymously into history, melted and remelted, diluted *ad infinitum* in solutions of African, European and New World gold. A few atoms of Atlantean gold must inevitably reside in modern wedding bands, or in the welding joints of the spent lunar module on Taurus-Littrow.

WINTER, 1,046,586 B.C.

On Thera, gold and silver were just beginning to percolate up through steam vents; but more than a million years would pass before anyone set foot upon the island and began collecting the treasure. The nearest people were across the southern Mediterranean, in Egypt and the Atlas Mountains. They were *Homo erectus*; and they had only recently (in terms of geologic time) ascended to ownership of the earth. A half million years earlier, *Homo erectus* had shared Africa, Turkey and Asia with other people, some brandishing stone hand axes and looking recognizably human, others astonishingly apelike in appearance. Most or all of them were gone, losers in nature's extinction lottery.

WINTER, 2,095,162 B.C.

The Ice Ages came and went in cycles of roughly ninety thousand years. Whole forests came, then vanished as drought swept across East Africa. Tool-making creatures came and did not go away.

We were *Homo habilis*, toolmakers who differed from *Homo erectus* in having a smaller brain and correspondingly lower forehead. Two-million-year-old *Homo habilis* bones, often associated with razor-sharp stone flakes, have been found scattered from Kenya to Java. Some members of this clan must have wandered into the isolation of northern Asia, because that is where the next branch in human evolution, *Homo erectus*, seems to have emerged. Once this happened, about 1,800,000 B.C., the new lineage spread across Turkey to the Mediterranean shore, and southward into Africa, where a major feature of *Homo habilis*'s environment became the presence of no fewer than three other manlike bipeds. Even before the arrival of *Homo erectus*, there were *Australopithecus robustus* and *Australopithecus boisei* to be dealt with.

Tanzania's Olduvai Gorge is replete with punctured shards of skulls. Louis Leakey and South African anthropologist Raymond Dart confirmed some of the holes as having been made by leopards' teeth. Other skulls, many others, bear marks of butchery with stone tools. In Kenya, pieces of stone tools intermingled with chopped bones and the remains of campfires suggest that Australopithecine was the flavor of choice.

We do not know precisely what role killing played in the ascent of man, but the picture is clear enough—chillingly clear. *Homo habilis*, the branch closest to *Homo erectus*, and hence its fiercest competitor, went first, ceasing to leave bones in the fossil record around 1,400,000 B.C. *Australopithecus robustus* followed less than a hundred thousand years later. Then, just short of one million B.C., *Australopithecus boisei* vanished. Man's first world war must have been a real dilly.

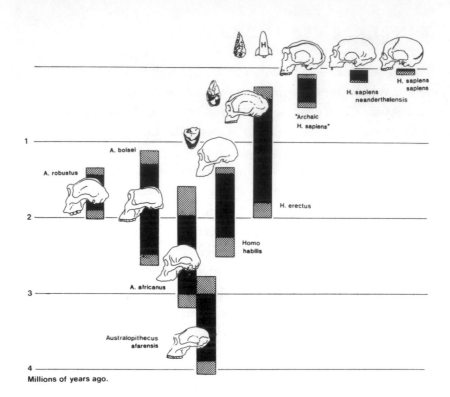

Branching lineages on the Australopithecus/Homo "tree." Illustration based upon skull reconstructions by Nicolis Amorosi.

WINTER, 4,192,314 B.C.

The African plate was driving Italy like a nail into southern Europe, pushing up the Alps. Greece, Crete and the Aegean islands were growing out of wrinkles in the earth, and in Ethiopia, where Arabia was being torn from Africa and the Red Sea was opening up, the earliest man-apes appeared.

We have pushed back, now, past the beginning of the archaeological record. The first stone tools turn up in strata dated to about 2.5 million years ago. This by no means fixes the backward limit of tool use. Even modern chimpanzees have been seen wielding sticks against each other. Not all of the stone tools in the world have been discovered, and—before they began shaping stones—our forebears almost certainly used softer, wooden tools such as sticks sharpened against coarse rocks. Stone tools are virtually indestructible; wood rots and seldom enters the fossil record, and because of this we may never know how far back the Wood Age extends beyond the Stone Age.

There is no telling whether "Lucy" or any of her contemporaries used wooden tools. Lucy was a young Australopithecine woman who died in Ethiopia about 3.5 million years ago. And that was that—at least until A.D. 1974, when Maurice Taeib of France's National Center for Scientific Research and Donald Johanson of the Cleveland Museum of Natural History brought her bones to light. Somehow, amid all the scientific bickering (sometimes leading to name-calling and food throwing) over Lucy's place in the history of life, she became a household name—which proves little except that in America even a fossil can aspire to stardom.

Bones very similar to Lucy's are found ranging from four million years ago to about three million years ago. During that time, her *A. afarensis* lineage seems to have undergone little or no evolutionary change. We know from the shape of her pelvis and leg bones, and from trails of fossil footprints, that she walked erect on two feet—but she stood somewhere between humans and chimpanzees and, judging from her skull, closer to chimps.

SUMMER, 8,386,258 B.C.

The Atlantic Ocean was narrower than it is today. Thera was eighty miles closer to Manhattan Island. The star at the tip of the Big Dipper's handle had shifted more than half way around the circle of the sky. Not one of our present-day constellations was even remotely familiar and the most distant galaxies were millions of light years closer to us than they would be in 1989. Whole islands of stars were (and still are) scattering away from each other, as if they were fragments caught in the aftermath of a stupendous explosion. If you could step back through deep time and bring radio telescopes with you, you'd be able to see that the temperature and density of the universe have begun to creep upward.

We know practically nothing about the evolution of our pre-human ancestors between four and eight million years ago. Digging down through the Pliocene Epoch, the spade strikes a wall near four million B.C. Beyond that point stretches Miocene time and very few hominid fossils (creatures with humanlike features). On our side of the wall stands Lucy, already walking erect as her ancestors had been doing for—how long? As Africa drifted north, carrying Lucy's furry forebears with it, a mountain range was pushed up that slowly bridged Spain and Morocco and put a natural dam across the Strait of Gibraltar. The Mediterranean Sea began to evaporate, leaving behind a desert basin two miles deep, and endless flats of salt. When eventually the Atlantic Ocean broke through, it did not so much dissolve the salt flats as bury them under layers of sediment. Drill samples collected during the *Glomar Challenger* expeditions reveal salt deposits underlying almost the entire Mediterranean. Caged within the salt crystals are fossil microscopic algae. Like all green plants they drew sustenance from sunlight, and their presence proves that shallow lakes once existed two miles down on the bed of the Mediterranean.

At that depth, the atmosphere piled up 1.5 times thicker than anywhere on earth today. The more richly oxygenated air would probably have been healthier for you; but dense air trapped more of the sun's energy, and temperatures on the Mediterranean basin must have been comparable to Upper Egypt on an August afternoon.

The basin was mostly desert. At the foothills of Crete, the Nile was the river of life, as it is on the desert above today.

When the Mediterranean first evaporated, the Nile spilled over a waterfall eight thousand feet high, about forty times higher than Niagara. The fall did not last—could not last. It chewed back the bare limestone, conveying broken bits of Egypt toward Crete and, as it went, forming great alluvial fans on the Mediterranean floor (they are still there today). It slashed the earth from Cairo to Aswan. When the Mediterranean rose again, the Nile silted the slash over; but oil geologists drilling through thousands of feet of mud have located the floor of Nile Canyon. It lies fifteen miles east of Cheops, a mile and a half beneath the city of Cairo.

Carrying stream-washed sediments from most of Africa, the Nile deposited fertile soil on the eastern end of the Mediterranean basin. It accumulated to a remarkable depth, hundreds of feet thick in places. Tall stands of palms sprouted on top of it, irrigated by the canyon to the south. Sheep-sized elephants, pygmy hippos and dog-sized deer whose fossils are unique to the Mediterranean islands might have originated on Mediterranean oases at this time.

North of the Nile oases were the Cretan and Theran highlands, which were arid at their foothills, but became relatively cool and well-watered higher up. They supported upland forests much like those growing six thousand feet up in the Atlas Mountains today. Into the Theran forests came migrating creatures from the basin oases. When the Mediterranean filled again, the migrants would be forever cut off from predators in southern Europe and North Africa, on mountaintops turned suddenly into Aegean islands. There, the bones of pygmy elephants and giant shrews would continue to enter the fossil record, until the arrival of seafaring humans near 6000 B.C.

WINTER, 16,775,226 B.C.

Thera was not. For tens of millions of years the island lay at the bottom of the sea, waiting to be born. The Mediterranean was an open body of water, connected at its eastern and western ends to the Atlantic and Indian oceans. This was changing. Africa was in collision with the lands of the Middle East. The continent's north-ward drift was pushing peninsulas and chains of islands on its eastern edge into their first physical contact with Europe. The animals living on either side of the junction would soon be meeting for the first time. Antelope and horses, which had evolved in Eurasia, would extend their range south into Africa, passing elephants, monkeys and apes coming from the opposite direction.

During this warmest part of the Miocene Epoch, lush forests advanced into Greenland and northern Alaska. Even the Antarctic coastline was green with plant life. Across the forests and savannas of Europe, East Africa and Asia, the cosmopolitan dryopithecine and sivapithecine "apes" were spreading. It was the zenith of apen radiation and diversification: a great branching of new lineages, most of which would die off during the next two or three million years. One line, probably a still undiscovered one, pointed in the direction of chimpanzees, gorillas and Australopithecines—beasts that became us. There was, in those days, nothing that even resembled humans. Protoapes were the newest and brightest things around.

The volcanic hot spot over which Thera would one day emerge was located 160 miles west of its 1989 position, on an earth that turned faster than it does today. Each year was shorter by almost six minutes. The general temperature of deep space was 2.82 degrees above absolute zero, 0.02 degrees hotter than it would be in 1989—the background echo of an event called the Big Bang. The density of the universe, too, was higher: at least three hydrogen atoms per ten cubic yards of space; in other words, about 50 percent more atoms than you would find today.

WINTER, 33,552,442 B.C.

We have ventured back to Oligocene time. The general temperature of the universe has risen to 2.84 degrees Kelvin. The world's climate, warming and cooling, time and time again, continues, in backward fashion, as a succession of steps up a staircase toward Cretaceous warmth. Continent-sized ice sheets have come and gone. We have watched human evolution in reverse until at last we can no longer recognize ourselves among the fossils. Thera has vanished. And yet, stepping back to 33,552,442 B.C., Split Rock has not moved an inch on Taurus-Littrow. It is the same shape, with the same cracks in the same places.

Africa, meanwhile, was hundreds of miles south of its A.D. 1989 position. The first true cats and dogs had appeared, and proto-monkeys lived among them in Egyptian forests. One branch of the feline tree was adopting an aquatic life, and would eventually beget seals.

Northern Egypt was under water. Clam shells were being slowly buried under the skeletons of microscopic animals called foraminifera. The calcium deposits had been piling up for millions of years, forming layers of limey mud that were in several places hundreds of feet deep. In time, the mud would harden into stone; and descendants of the spidery primates that roamed Africa's shores would one day cut the stone into huge blocks, carry them away on barges, and build from them the Great Pyramid of Cheops at Giza.

AUTUMN, 67,016,874 B.C.

We have just stepped back across 33.6 million years, during which time the earth has changed more than in all the previous time jumps combined. Paleontologist Gerard R. Case has identified about fifteen species of fossil sharks dating to late Cretaceous time, each of which has left traces of itself (mostly teeth) in Morocco's Atlas Mountains and under the fertile Marlboro plains of New Jersey. This makes sense, in light of the fact that the Atlantic Ocean was then only half its present width.

Above unborn Thera, the Mediterranean was an open ocean that rivaled the Atlantic. Dinosaurs swam in it. They roamed the continents, ascended mountain peaks, penetrated into Alaska, grew feathers and fur. *Seismosaurus* (so named because it must have been detectable on seismographs when it walked), was ten stories tall. Other saurians were as small as mice. Some were probably as active and warm-blooded as birds and bats. Others, like the foot-long tuatara (which survives today in New Zealand), were as cold-blooded as crocodiles, while still others were somewhere in between.

The most memorable qualities of the saurians are that some of them were very big and that, like Plato's Atlantis, they vanished. Much like the Minoans, whose civilization might already have been in trouble, but who finally lost a lottery when Thera exploded, the saurians, and most of the species that shared the world with them, had been dying off for five million years, and finally collapsed into extinction when an asteroid struck the Earth.*

If not for five million years of deteriorating climate and an asteroid, pterosaurs might now be surveying Thera from on high. If not for the Theran upheaval, Minoans might now be exploring Alkaid. Without the asteroid and Thera, highly evolved saurians could be living in cities along the Nile, and hurling space probes at Jupiter, and we humans might not exist at all. Paleontologist Stephen Jay Gould compares this scenario with the movie classic *It's a Wonderful Life*, in which an angel shows George Bailey (James Stewart) how his community would have been altered (for the worse) if he'd never lived.

"We are here because events happened to take particular, random turns," says Gould, "not as a result of a tendency for nature to lead, step by sequential step, up a straight and narrow ladder from intertidal slime to man and civilization. Just as the choices George Bailey had in life seemed insignificant to him, yet in the long run made life in his community completely different, so too can a random volcanic explosion, or a bit of celestial buckshot,

---

* At least one asteroid (probably two of them, associated with an increase in the cosmic dust background) left fingerprints in the geologic record, primarily in the form of a worldwide layer of platinum group elements (rare on the earth's surface, abundant in cosmic dust, meteorites and asteroids) mixed with microshards of squashed quartz.

send life on earth in wholly new directions. Such a past gives us the freedom—and responsibility—to choose our future."

SUMMER, 134,215,738 B.C.

South America was welded solidly to Africa. Europe and North America were joined at Spain, and the Mediterranean was the world's second largest ocean. If you had a ship, you could sail east from Morocco, over the place where Thera would one day stand, and straight into Pacific waters.

We have retreated deep within the Mesozoic Era—to the early part of the Lower Cretaceous, to be precise. Dinosaurs have been with us for fully one half of our backward journey, and they go back farther still. They were not merely peculiar beasts that happened to live for a while and then died. They were the dominant life form of an entire planet for more than a hundred million years; and the mammals that have dominated the upper 1.3 percent of earth history did not simply appear at the end of the Dinosaur Age. They appeared with the dinosaurs, but could not displace them. In 134,215,738 B.C., our cynodont forebears had been leaving behind fossils for sixty million years. Most were egg-laying carnivores. All of them were small, furry and ratlike. None of them were particularly common. They lived forever under foot in the nooks and crannies of a world whose chief denizens had scales.

The extinction of dinosaurs and other species is what gives the earth its history. If all species had been born with the earth, on the same day, and lasted forever, we would have no perspective, no sense of the arrow of time. Digging back hundreds of millions of years, everything would be the same. You'd find Neanderthals among Cretaceous dinosaurs, brontosaurs among Minoans. But as the tyrannosaurs and brontosaurs disappear and Australopithecines or pygmy elephants take their place, we begin to see the stages of a lengthy history, a sequence in which the farther back we go, the more unlike modern creatures the animals become.

SUMMER, 268,433,466 B.C.

Taurus-Littrow looked very much the same; but you would not have recognized the planet below. The Mediterranean was wider than today's Atlantic, and the Atlantic did not exist as yet. Thera (or at least the strip of land over which Thera would one day exist) was part of a vast northern hemisphere continent called Laurasia. Africa was huddled at the South Pole with South America, India, Australia and Antarctica. They were now the continent of Gondwanaland. During the next fifty million years, Laurasia and Gondwanaland would coalesce to form the supercontinent Pangea, and a whole hemisphere of the earth would become unbroken ocean.

In shallow seas and swampy lowlands, the only fossil samples of life from this period were forming. In Africa and South America—now jammed together—the little Permian reptile *Mesosaurus* harassed bivalved, shrimplike animals called branchiopods. While other vertebrate branches were adapting to life on the continent, *Mesosaurus*'s ancestors had already been there, and at least one branch was on its way back to sea.

To comprehend time, even when you are descending through it by powers of two, you should step back and take a grandstand view now and again. When Thera vanishes in sixteen million years and migrates across a shrinking Atlantic to become part of Laurasia, what has happened? In sixteen million years a volcano grows thousands of feet high (with rain and gravity, every inch of the way, trying to tear it down), and drifts one hundred sixty miles on the globe. Each million years it drifts ten miles, which means each thousand years it drifts barely sixty feet. The drift per year is detectable only with laser beams, and few people on earth are aware that it is happening.

The "palace" at Knossos was home to a dynasty of kings named Minos. It appears to have served as a cultural and industrial center, and contained warehouses of man-sized storage jars with a capacity exceeding sixty thousand gallons. Apartments in the capital city stood up to six stories tall and were provided with plumbing systems that would not be duplicated anywhere on Earth for nearly three thousand years. Some of the homes were actually equipped with showers.

C. R. Pellegrino

*T*hera Lagoon is actually an eight-mile-wide blow hole in the earth, created in the span of a deep sigh. That quickly a mile-high mountain had disappeared, and a mile-deep hole stood in its place. For a time, walls of water hovered outside the hole, then spilled in to create waterfalls like none the world had ever seen. The ship seen in this photomosaic, scaled against the Theran Falls, would have seemed no more significant than a log flung into Niagara. In the center of the lagoon, a new volcanic cone, called the "New Burnt Land," has been slowly rising since the explosion of 1628 B.C. The village of Fira is seen in the extreme left. Fira Quarry stands just behind it. *Photomosaic by C. R. Pellegrino*

A close-up view of the "New Burnt Land," one of the most desolate places on Earth, located in the throat of the planet's deadliest volcano.
C. R. Pellegrino

*I*n the year 1630 B.C., West House had stood at least three stories tall. It would not have looked out of place on a modern Theran street, and was even equipped with sophisticated plumbing, including a bath and a flush toilet. The city within the earth is covered over by a tin roof, which ascends to

ground level and protects the buildings from wind and rain (without the roof, the entire city would turn to dust very quickly). While excavating shafts for roof supports, workers discovered a second city beneath the buried city. It is two thousand years older. *Photomosaic by C. R. Pellegrino*

A view of the back yard of Delta House, showing "fossil" wooden beams and pottery jars turned upright sometime between the first major earthquake and the final evacuation of the island near 1630 B.C..

C. R. Pellegrino

Thirty-six hundred years have oxidized and disintegrated all the organic furnishings of the buried city. The wooden beams shown here were merely hollow spaces in volcanic ash when archaeologists found them. Before digging down to street level, they pumped concrete into the voids and reinforced the liquid with steel wire. When the excavators removed the ash, the hardened cement had assumed the shape of the vanished wood, and supported the stone blocks as the wood once had. Even the original wood grain can still be seen in the concrete "fossils." *C. R. Pellegrino*

(Below)

One of the buried rooms in the Delta Complex depicts beautifully stylized antelopes at play. The antelopes are painted with strokes reminiscent of (though presaging by nearly 4,000 years) Pablo Picasso—yet they are, at the same time, so true to life that they can be identified as a species found only in eastern Africa near the headwaters of the Nile, suggesting that these prehistoric people, at least, traveled farther than most might think. *Otis Imboden/© National Geographic Society*

T *(Left)*
*T*elchines Road meanders through the center of the excavation. Spyridon Marinatos, who discovered the buried city and began his dig near this very spot, died here, and is entombed in Room Delta 16 at right. *C. R. Pellegrino*

T *(Below)*
*T*he Room of the Lilies in Delta House was strewn with vessels and furnishings including a rattan bed, whose fossil impression is revealed here in a plaster cast. Evidently, objects were piled haphazardly into this room during the city's final abandonment. Behind the fossilized bed frame, a wall mural depicts a hybrid of the white madonna lily and the red lily. The flowers grow atop volcanic boulders, nodding as if bent by a gentle breeze, while swallows dart above them. The fresco expresses Theran spring before the upheaval of 1682 B.C. Though swallows inhabit all the nearby islands, they never did return to Thera. "Perhaps the eruption was so terrifying as to uproot even their migrating habits," suggested Spyridon Marinatos. "A race memory, deep-wired and instinctive: Stay away from Thera!" *Otis Imboden/*
© *National Geographic Society*

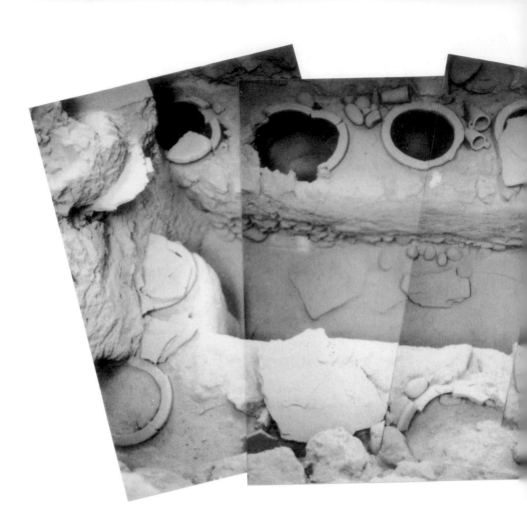

*A*n abandoned "bake shop" at Delta House has lain undisturbed for 3600 years. There is something hauntingly beautiful yet at the same time heartbreaking when you look down upon ceramic scoops, storage jars and bits of grain that might have been left in their present positions only yesterday; and then you understand that this civilization is now so utterly gone that no one remembers its name. *Photomosaic by C. R. Pellegrino*

A powerful earthquake preceded the final abandonment and burial of Thera's lost city. In one home, a stone stairway was cracked in half. *C. R. Pellegrino*

$\mathcal{A}$ *(Left)*
t the very rim of the Thera crater, one of the world's most infamous donkey trails follows a zig-zag path to the cliff-top village of Fira. This pair of photos was taken by Arthur C. Clarke during his 1965 visit to the island with rocket pioneer Wernher von Braun. A popular myth has it that Thera's donkeys are inhabited by souls sent from purgatory. Burdened with passengers and produce, they pay penance for their sins by climbing up and down the thousand-foot cliff twenty times a day. On other Greek islands, and as far away as Egypt and Syria, donkey owners warn stubborn or rebellious animals, "Behave or I'll send you to Thera!" *Arthur C. Clarke*

$\mathcal{S}$ pyridon Marinatos (facing) and a fellow archaeologist (foreground) crouch in a tunnel that, during 1967, followed Telchines Road into the lost world of the Minoans. Marinatos was only the first in a long dynasty of archaeologists who would require an estimated three hundred years to completely excavate the buried city. *Otis Imboden/© National Geographic Society*

"*I* was seventeen years old when my father took me to the site for the first time," recalls archaeologist Nanno Marinatos. "There were rows of man-sized pottery jars emerging from the white ash, still standing upright in exactly the positions they had been left some thirty-six centuries ago. Chemical analyses would soon show some of the vessels to contain the dried remnants of wine and oil. Others were filled with fossilized grain, and another with the shells of landsnails. 'A Bronze Age feast!' my father observed." *C. R. Pellegrino*

*M*any centuries after the Thera upheaval, Egypt and Rome claimed the island as a naval port. The Romans built an amphitheater atop Mesa Vouno. Here the drama was music, dancing and riotous skits. *C. R. Pellegrino*

*T*wo hundred feet of volcanic ash lie on top of the buried city, and on top of the ash stand vineyards, churches and the little farming village of Akrotiri. *C. R. Pellegrino*

Imagine you are having lunch in one of Fira's cliff-side restaurants overlooking Thera Lagoon. Below you, donkeys labor up hazardous roads that ascend almost one thousand feet above sea level. In 2106 B.C., those donkeys would have been embedded in the solid rock of a volcanic peak rising higher than Fira and occupying the whole lagoon. The island had hot springs then, and cities, and bull arenas. Six million years earlier, Thera towered over baking Aegean salt flats. It was capped with an oasis and supported giant shrews. Sixty million years before that, plesiosaurs swam over the site; and two hundred million years before that, club mosses grew over the same spot, grew as tall as cedars. Something resembling a two-ton salamander waddled on its belly, scratching a permanent track across a Coal Age mudflat. Five random snapshots from the upper six percent of earth's history.

SUMMER, 536,868,912 B.C.

The universe was about 0.5×10^{-29} as dense as water, meaning that every cubic yard of space, no matter how far away it happened to be located from planets, or stars, or galaxies, contained at least ten atoms. These were heated to at least 3.43°K. Put another way, cold empty space was a hundred times as dense as it is today, and fifty percent warmer.

On earth, these were middle Cambrian times. Antarctica and Canada were located on the Equator, North Africa and Brazil poked upside down into the South Pole, and there is almost no telling where Thera was. Save for a few wind-blown knots of DNA (virus particles, if such existed), dried bacteria, algae and protozoans, the continents were probably without life.

Virtually the only fossils we know from this time come from the Rocky Mountains of British Columbia, near a place called Burgess Pass, where an algae-secreted reef once grew. It rose vertically more than three hundred feet from the sea floor and formed a front many miles across.

At the base of the reef fine silts were gathering. The steady buildup of mud triggered frequent avalanches that often spilled from the reef crest as gigantic, rolling clouds, carrying with them any creatures that lived in their path and burying them in deeper waters beyond the reach of both oxygen and scavengers. As the mud gradually compacted and consolidated into shale, the buried animals were flattened and their soft parts became thin stains of feldspar. Some of the fossils are so strange that their discoverers, equating them to something out of a peyote nightmare, have given them such names as *Hallucigenia*—which describes a thing about four inches long that drove along the sea floor on seven pairs of sharply pointed, stiltlike legs. Seven tentacles ran the length of its back, each ending, apparently, with a mouth.

In the waters near Burgess Pass, and in the world outside it, there were varieties of life that we will never know from fossils. In one little corner of the earth, conditions were just right to preserve a large number of soft-bodied animals for a half billion years; but even here we can have only the most fleeting glimpse of what actually existed.

It is a sad fact, but a fact nonetheless, that as we accelerate away from time present the world becomes more alien, more interesting and far less knowable.

SPRING, 1,073,739,824 B.C.

The place where Thera would one day stand was under water. In Canada, bits of seaweed began to enter the fossil record. The seas were dominated by very simple organisms, and many of them were unicellular. Some used energy from the sun to process carbon from the environment. Others, crowding around submarine vents, used sulfides instead of sunlight. Others devoured their neighbors, and still others belonged to enormous colonies resembling filaments, balls and medusae. It is possible that several lineages of wormlike creatures had evolved by this time. If so, they swam on the edges of continents that must have been shuffled as much during the interval between 1.07 billion B.C. and middle Cambrian time as during the succeeding 537 million years.

SPRING, 2,147,481,648 B.C.

Cyanobacteria (blue-green algae), protozoans and green multi-cellular threads rioted in the oceans. Some of the bacteria grew to form huge reefs called stromatolites, and as the reefs grew, they consumed carbon dioxide and released oxygen. Up to this time, free oxygen had been virtually absent in the atmosphere. Now it was accumulating on a grand scale—actually on the verge of becoming poisonous to life on earth. The oxygen crisis, generated by life itself, became history's first major natural selection filter. Organisms were either able to withstand oxygen exposure, or possessed cellular machinery that could be adapted to its use as a fuel source, or they died. We know that the oxygen crisis existed because rocks from this time tell us that free oxygen had begun to combine with and precipitate iron from sea water. More than ninety percent of the world's iron deposits were laid down between 2.5 and 1.8 billion B.C. Seen from Taurus-Littrow (which still looked very much as it does today) the earth was blue ocean, swirls of white clouds, and continents as red as the sands of Mars.

The oxygen also liberated uranium compounds into the water. Bacterial reefs stockpiled them in ores, and one reef, in what is now western Africa, concentrated so much uranium-235 as to become the world's first known nuclear reactor. The bacteria burned to death in a nuclear storm of their own creation.

SPRING, 4,294,965,296 B.C.

It was hot.

The earth was practically new, a mere three hundred million years old, give or take a few tens of millions of years. Much of the planet was still steaming from the heat of accretion and from the decay of short-lived radioactive elements injected into the solar system almost at the moment of the earth's formation.

From continent-spanning nets of wrinkles and ruptures emerged outflows of basaltic lava wide enough to accommodate the Aegean Sea. There were steam and dust and radioactivity in the air, and torrential rains. No splashes of green challenged the forces of erosion. Almost as quickly as it formed, the bare bedrock was hewn down into gravel and powder. Where the seas—actually little more than oversized lakes—met land, the beaches were glittering black bands of volcanic sand.

University of Miami microbiologist Sidney Fox has provided a laboratory analogy for what probably happened next. He creates organic microspheres by first stirring amino acids into water and then evaporating the water by pouring the mixture over hot volcanic rocks. Finally, he splashes water over the mixture, and the resulting runoff is full of protein microspheres. You can cook up such structures in your own kitchen in about twenty hours. Simply take any combination of substances containing carbon, hydrogen, oxygen and nitrogen, and apply any form of energy— electricity, heat, shock waves—then shake the resulting mixture of protein building-blocks vigorously, and you will come up with microscopic bags and tubes of protein. They are not life (though some of them will scavenge more protein, grow and divide), and they were not life 4.3 billion years ago, but they were getting close to it.

Our bodies ripple with energy-releasing and energy-capturing molecules that conduct symphonies written on DNA and performed by protein. The music might have begun, at one time or another, in the exhalations and regurgitations of volcanoes. It begins to look as if the composer in our lives is the earth itself, and that the poets were right all along. We live, every one of us, on the surface of a parent, not just a planet.

WINTER, 8,589,932,592 B.C.

It was all out there somewhere—all of it amidst the stars. There went Theran gold, there went the iron that runs in your veins, there went calcium from the bull's horn, and carbon from the pages of this book. The earth and the moon and the stars of the Big Dipper did not exist as yet; and if we double our backstep again to 17,179,867,184 B.C., there is almost no telling what will and will not be. What *can* be told is that if we take the A.D. 1989 rates at which galaxies both near and far appear to be receding through expanding space-time and clock them back as far as we can, we find that they all come crashing in on our heads between twelve and seventeen billion B.C. Almost nothing is known of this strange, cosmic crunch period. The universe was simply very dense and very hot. What little we do know is glimpsed in a few very special atomic accelerators, where matter and antimatter can be collided to create, for a few trillionths of a second, temperatures and pressures close to those found in the crunch. Unless there are other electronic civilizations amidst the stars, the hottest places in the universe are of human origin.

Probing back to Time Zero plus fourteen seconds, we encounter temperatures exceeding five billion degrees. A second earlier it is hotter yet—so hot that electrons, positrons (the positively charged antimatter counterparts of electrons) and photons of light are blasted to scattered wave fronts as soon as they appear.

At Time Zero plus $\frac{1}{10,000}$ of a second, the temperature has risen to degrees measured only in the unimaginable realm of millions of trillions, and the space today occupied by a virus is denser than the earth, denser than the earth and the planets of the solar system and the sun put together. During this first chip of time, everything you can see—all the stars of the Milky Way, Andromeda and the most distant galaxies—would fit inside the diameter of your thumbnail.

When the universe is 10^{-43} of a second old—when its density is 10^{94} times that of water—we smash into a rather nasty thing called the Planck Barrier. It is the astrophysicists' version of the old mapmakers' "Here Be Monsters." Beyond the Planck Barrier lies chaos and the inevitable breakdown of known physical laws. Princeton's John Wheeler can be credited (or blamed) for discovering the Planck Barrier; it translates literally to the other side of time. He visualizes Planck space-time as being hopelessly knotted up in a turbulent foam. Temperature and density soar to the infinite, pulling the "here" and "now" and "then" apart until we can push back no farther. Time is no longer understandable in terms of orderly, sequential events. If you've ever wondered what life is like in a black hole, drop in at the Planck Barrier and look around.

WINTER, A.D. 1989

So it looks as if the universe rose from infinitely curved, infinitely dense and infinitely hot space-time. And where did that come from? Good question.

Our present knowledge suggests that there is enough mass out there—which seems to be holding the galaxies in tight clusters—to eventually pull the universe back from its current expansion into another cosmic crunch, to create another Big Bang.

Perhaps the first Big Bang was not the beginning at all, but the end of a previous ordered phase—another universe. Perhaps all of our history, through the dinosaurs and the Minoans to the present, is (as Carl Sagan once suggested to me) merely the most recent cusp in an infinite series of cosmic expansions and contractions.

And from this line of thought arises a strange, haunting and evocative possibility: could an infinite series of rebounding universes be connected somewhere, like a snake head to tail? Can it be that baby universes actually bud off from each other, living and dying in what might just as well be a jiffy?* Could our present rebound, expansion and eventual collapse be identical to one that has gone before it, and another that will come after? Every atom may ultimately wind up in the same place all over again, every grain of dust in this room, exactly where it is today, over and over again. The children of Poseidon will settle the high-risk real estate round Thera, repeating the same mistake, and billions of years later carbon from your fingertips and carbon from an Atlantean grave will plunge side by side into the next cosmic crunch, into the ultimate black hole, where time itself becomes timeless.

Untold billions of years ago, and untold billions of years from now, you might have sat in this same place—and you might sit in this same place—reading these same words.

---

* A jiffy is the travel time of light across the diameter of a proton, or one billion-trillionth of a second. (Hence the phrase, "Be back in a jiffy.")

We look back through countless millions of years and see the great
will to live struggling out of the intertidal slime . . . we watch it draw
near and more akin to us, expanding, elaborating itself, pursuing its
relentless inconceivable purpose, until at least it reaches us and its
being beats through our brains and arteries. It is possible to believe
that all that the human mind has ever accomplished is but the dream
before the awakening. A day will come, one day in the unending
succession of days, when beings, beings who are now only latent in
our thoughts and hidden in our loins, shall stand upon this Earth
as one stands upon a footstool, and shall laugh and reach out amidst
the stars.

—H.G. Wells, *An Experiment in Autobiography*

There are no men born ahead of their times. There are only visionaries, and there are circumstances; everything else is hindsight.

—Pellegrino's second law

What seest thou else
In the dark backward and abysm of time?

—William Shakespeare, *The Tempest*

THE
THREE-POUND
TIME
MACHINE

SUMMER, TIME PRESENT

Cooked Australopithecine bones unearthed in a South African cave suggest that *Homo erectus* began controlling fire somewhere between 1 million and 1.5 million years ago. Presumably, language and human consciousness had also dawned by that time.

If the *Homo erectus* brain ever did question its origins, it had no knowledge of where the rocks had come from or when they had formed, no yardstick by which to measure or understand the past. Deep time, from *Homo erectus*'s perspective, probably went back past one pair of parents to a grandparent or two; and there, a very long time ago, lay the beginning of all things.

We have a grander perspective. We can look back past *Homo erectus* to dinosaurs, and past them to the first chip of time, when a singularity began to expand at the speed of light, and photons of light emerged, all at once, from every nook and cranny in the universe. From this beginning came particles now assembled into retinal cones and the columns of the neocortex.

And when we think and move, what really happens? As you read, you are shuffling electrons whose existence goes back before the earth, before the dawn of brains and life, before the first stars. Pulses of electrons are directed through our brains in an organized fashion. They are the basis of every thought we have. But imagine, for a moment, what you would look like if all the organic molecules of your body were made invisible, so that you could look in a mirror and see only the paths of freely moving electrons. Your outline would be there in every detail, brightest at the brain and spine: even the nerves in your fingertips and eyelids would show up as streams of electrons. Perhaps it is really the electrons who are reading these words. Perhaps our bodies are little more than vessels serving their interests.

Homo erectus; the rise and fall of civilizations; robot spacecraft and the grandmasters of computer chess (and what is the computer revolution but man's creation of a newer, perhaps better electron vessel?): all of these things have led up to a day when electrons, racing through the biggest, most sophisticated brains that have ever existed on the planet, could look back along the stream of time and ask, "Where did we come from?"

The electrons coursing through your brain are billions of years old. Compared to them, a million years is a short time. It did not take very long for a small, isolated tribe of *Homo erectus* to become archaic *Homo sapiens*; for archaic *Homo sapiens* to radiate out of their place of origin into Eurasia, Africa and Australia, eradicating *Homo erectus* and the Neanderthal tribes as they went; for their descendants to build the Minoan civilization, and for the elder civilization, as it fell, to disperse its knowledge to Greek successors, as Plato told us they did in the *Critias*.

Equipped with brains weighing roughly three pounds, we have started to appreciate the time experience of the electron. We can, with a little practice, begin turning our minds to archaeological, biologic and geologic time scales. These are the planet's time scales, and a true feeling for them creates a wonderful sense of companionship with the earth.

Your mind is a time machine, of sorts. Of the countless million forebears whose DNA runs in your veins, you are among the very first to live outside the present. Man is the only species known to arrange the events of its life into deliberate, constant intervals of time. But this goes beyond the simple measure of years A.D. or B.C., or vivid recollections of what you were doing a week ago today. Out of experience of cause and effect, we learn to anticipate the future in reasonable detail.

It is likely that human beings have been doing this for more than a half million years. One of its earliest consequences, as demonstrated by Neanderthal graves suggesting a belief in the existence of immortal spirits, must have been an awareness that death was not merely something that happened to everybody else. Hand in hand with this knowledge comes the realization that the stream of time neither begins with one's own birth nor ends with one's own death. Instead, it runs away from us in opposite directions—past and future.

SUMMER, A.D. 1965

The time travelers had come to Thera.

High atop Mesa Vouno, the historian Walter Lord explored the ruins of an unusually small amphitheater. His mind slipped easily into the past, seeking out the amphitheater when its limestone columns were unbroken and freshly painted. He saw actors— amateurs mostly, probably from the soldiers' own ranks—performing riotous skits, relieving as best they could the monotony and isolation of an Egyptian naval outpost.

The mountaintop fortification dated back to the time of Julius Caesar and Cleopatra. It was a popular archaeological site on the island—it was the only archaeological site. Yet at its feet, only a short distance away, near the little fishing and farming village of Akrotiri, lay a buried city of which the historian could only have dreamed in his wildest imaginings. It was haunting and beautiful, and nearly two thousand years older than the amphitheater; but Walter Lord would never see it. He was returning to America in a day or two, and only now was the city about to be discovered.

Five miles away at Thera Lagoon, two other time travelers had arrived. They were Wernher von Braun and Arthur C. Clarke. Unlike the historian, their minds tended to move forward in the stream of time, toward the future. The same species that once built amphitheaters on mountaintops was now on the verge of breaking away from the earth it had inhabited for a million years; and von Braun was one of the leading visionaries behind the escape plan. It was a long way from the atmosphere-piercing rocket-bombs of World War II to the 363-foot-tall Apollo moonship, yet both operated on the same principles, and could trace their ancestry directly to the designs of the late Robert Goddard.

During the 1930s, the spreading German empire had seen rockets as a niche (or legal loophole) not covered in the ban on German rearmament imposed by the Treaty of Versailles. While Goddard continued, in America, to work with an almost nonexistent budget, using parachutes hand-sewn by his wife—and nurturing his reputation as a "crackpot" in the process—the Germans studied his patents, and began pouring virtually unlimited funds into the enlargement of his designs. But von Braun's supervisors saw immediately that he was more interested in designing moonships and space stations than in bombing London from space. Before the technological horrors of World War II were over, he was imprisoned, declared an "unperson" and scheduled to be shipped away to "permanent rehabilitation" in a concentration camp. The only thing that spared him was the unavoidable fact that, without the dreamer, the rocket program was over. By 1945, the von Braun team was at work on the world's first reusable space shuttle. By 1955, as an American "prisoner of peace," von Braun worked on nothing much at all. The budget for rocketry and space exploration was nonexistent. So was the U.S.-Soviet space race. So was

NASA. Now, in keeping with the wishes of a martyred president, he was to send men to the moon by the end of the decade.

Clarke could claim some credit for making von Braun's dreams real. He had gained only small notoriety through the appearance of his 1951 book, *The Exploration of Space*, but ten years after publication—and in a way he could never have anticipated—the effort of writing it was justified. A copy had found its way into the hands of President Kennedy, and fired his imagination. Now, lunar modules were being built, and a man named Stanley Kubrick had approached Clarke to suggest that they team up to create "the proverbial good science-fiction movie." He was talking about *2001: A Space Odyssey*.

The arrival of Arthur C. Clarke and Wernher von Braun on the island at this time was a classic example of serendipity (a word that is derived from one of the many ancient names of Clarke's island residence, Sri Lanka). They had been attending the International Astronautical Federation Congress when a local Greek shipowner invited twenty delegates on a weekend cruise. Neither Clarke nor von Braun had ever heard of Professor Marinatos's suggestion that Thera and the Atlantis legend might be connected. Indeed, they'd never heard of the island until the day they arrived.*

On donkeyback, they climbed a zigzag road up Thera's eastern wall of rock. It was an uncomfortable ride. When the animals weren't trying to scrape you off against the wall, they were walking briskly along the very edge of the road, only inches away

---

* During this visit, Clarke persuaded the Athens Museum to let him examine the two-thousand-year-old Antikythera Mechanism—another of the great "what-ifs" of history. Discovered in a shipwreck, and believed to have originated somewhere between Crete and Rhodes, the device contained thirty-nine gears arranged in layers. Three dials displayed a read-out of the cycle of eclipses, the phases of the moon and the sun's position in the Greek zodiac for every day of the year—past, present and future. It even included a differential gear shift, permitting two shafts to rotate at variant speeds, just like the gear shifts developed for automobiles of the twentieth century. Someone had built an analog computer.

from a sheer drop that few tourists dared approach on two feet. After several hair-raising moments in which Clarke imagined himself about to ride off the face of the cliff, out into empty space, he decided that the only excitement in a donkey's life must be to scare the daylights out of unsuspecting tourists, and that he was not really riding a brain-damaged bag of bones intent on sacrificing itself to gravity.

Four hundred feet above the lagoon, they ascended through alternating bands of white, red and black lava. Despite the donkeys, their attention was drawn irresistibly to the rocks. When the mountain exploded in 1628 B.C. it opened up a time portal, a slice through the earth's skin.

Keeping both hands on the reins, von Braun motioned with his head. "Thousands of years here. Thousands of years to form just these few feet of bands."

"And that was only part of the story," said Clarke.

As indeed it was. The colored layers were clues to events that had formed the surface rocks of Thera. The layers were successive lava flows, and on each of them, in its turn, trees and grasses had grown. The youngest of them had formed a long time ago—certainly long before the last major ice advance began. They probably went back past the dawn of man: directly overhead lay six hundred feet and hundreds of thousands of years' worth of rock, and blanketing that was a layer of purest white. Seen from a distance, it looked as if the Theran cliffs were covered with deep snow. The snow was volcanic pumice, and it represented the final layer, most of it deposited in a single day more than thirty-six hundred years ago, when Thera Lagoon was carved out and the portal opened up. If you looked carefully at the snow, you'd see that there were shapes on top of it. Theran pumice and limestone could be combined to make a particularly strong cement, and from these materials the white-walled town of Fira had grown.

The fourth Theran time traveler was not aware of the other three, nor the other three of him—which was a shame, because all four would have loved to meet and spend an evening exchanging visions of past and future over octopus and wine.

The fourth traveler was Spyridon Marinatos. Even Thera's overall shape communicated hints of Plato's Atlantis story (*Critias*, 113d–e) to him. The mythical continent was said to have been a small island when the god Poseidon landed there and fell in love with the mortal Cleito. He moved over the face of the waters and a wave rippled out in circles from the island, making alternate zones of sea and land. The original, central island was encircled by a lagoon whose farthest reaches were the inward shores of a ring-shaped island. This island was ringed in turn by a third island, so that an alternating pattern of sea to land to sea again made it impossible for men in ships to reach the central island and disturb the two lovers. Wheels within wheels within wheels: the children of Poseidon and Cleito went out from the axis and became the inhabitants and rulers of Atlantis. Successive generations are said to have built a royal palace in the center, where the god and their ancestors had resided.

"This they continued to ornament in successive generations," wrote Plato (*Critias*, 115c–d), "every king surpassing the one who came before him to the utmost of his power, until they made the building a marvel to behold for size and beauty. . . . The stone which was used in the work they quarried from underneath the center of the island and from underneath the zones of land on the outer as well as the inner side. One kind of stone was white, another black, and a third red. Some of their buildings were simple, but in others they put together different stones, varying the color to please the eye, and to be a natural source of delight . . . They constructed bridges over the zones of the sea which surrounded the ancient metropolis, and made a passage into and out of the royal palace. . . . And, beginning from the sea, they dug a canal three hundred feet in width and one hundred feet in depth . . . which they carried through the outermost zone of land [hundreds of miles], making a passage from the sea up to this, which became a harbor, and leaving an opening sufficient to enable the largest vessels to find ingress."

Then, according to the legend, Atlantis sank, but not completely. The Egyptian priests told Solon that there were, remaining in small islets, the bones of the wasted body. The skeleton of the country was still out there, somewhere in the far west.

Marinatos suspected that, centuries before Solon's visit, Egyptian boats must have sailed into the smoldering wreck of Thera Lagoon. They might even have been drawn there by tales of a lost civilization and a mighty detonation in the west. The explorers would have found the remnant of a mountain smoking and booming in a circle of water whose circumference was a wall of freshly broken rocks. If Thera Lagoon was taken to be a mere fossil of what had once existed, it was easy to imagine openings through the cliffs—through which whole fleets of ships could now pass— as earthquake-shattered vestiges of canals. Ripped open, strewn about and buried, the land displayed no sign of ever having been inhabited, though it was, within recent memory, a thriving metropolis. If palaces and cities were missing and presumed sunk, why not bridges between two zones of land? And why not a third, outer zone of land? Or a fourth, large enough, perhaps, to encompass crescent-shaped Crete?

It was possible, Marinatos decided, that the shape of Thera was one of several grains of truth in Plato's tale. No other island in or near the Mediterranean exhibited Thera's bull's-eye configuration. The description of Atlantis as concentric rings of land and water must have been a remote and poorly conceived extrapolation from Thera's posteruption geography.

If the tale Solon heard in Egypt was a genuine though not clearly understood memory of Minoan Thera and Crete, there is no indication that he knew it. To him, the story was merely grist for his mill. He had an epic poem in mind, which, according to Plato, he began outlining before his death, substituting Greek names for Egyptian ones (Atlantis for Keftiu?), and adding historical and geographical details. Had Solon finished the poem, he would today be as well remembered as is Homer for his *Iliad* and *Odyssey*. But he died, and the unfinished manuscript was handed down through his family. Nearly two hundred years later Plato realized the dramatic possibilities of the story, and began his own version. He, too, died before he could finish. The Atlantis manuscript ends tantalizingly in midsentence:

> Zeus, the god of gods, who rules according to law, and is able to see into such things, perceiving that an honorable race was in a woeful plight, and wanting to inflict punishment on them, that they might be chastened and improved, collected the gods into their most holy habitation, which, being placed in the center of the world, beholds all created things. And when he had called them together he spake as follows . . . (*Critias*, 121c)

Marinatos supposed that the enlarged size of the island, its dislocation outside the Pillars of Hercules (where the Greeks had a tendency to place ancient, mythical events) and other embellishments were added by Plato: bridges and Egyptian canals sprang up, then grew out of all rational proportion; and the palace architecture was gilded with details of Far-Eastern splendor borrowed from the Greek historian Herodotus. It seemed that a few touches from Homer's description of Phaeacia had also crept in— which was a curious resonance, as Phaeacia was itself a distant memory of Minoan Crete.

And then there were the elephants. Egyptian wall paintings accompanied by hieroglyphic references to Keftiu showed men wearing Minoan kilts identical to those depicted on Cretan frescoes. The men were also bearing, as gifts, what looked for all the world like baby elephants' tusks. Marinatos had never seen the wall paintings, and he could not know that pygmy elephants had actually lived on the islands, or that early Minoan settlers probably hunted them to extinction. No paleontologist had yet discovered their bones, so Marinatos dismissed Plato's mention of elephant herds on Atlantis as just another embellishment.

"Atlantis was the way to other islands," Plato had said (*Timaeus*, 24e). Notably, Plato's Atlantis seemed, at times, to be just an island. "And from these you might pass to the whole of the opposite continent." From an Egyptian point of view, this seemed to Marinatos an apt description of Crete: a gateway to the other Minoan islands (the Cyclades) and the European continent beyond. The size and location of Atlantis's capital was also more in proportion with Crete or Thera than with Plato's Libya/Asia-sized continent. Plato's mythical palace was located halfway along the coast of the island, about fifty *stadia* (a stadia being less than half a mile) inland, on a low hill about four stadia across (*Critias*, 113). The real Minoan palace of Knossos was located halfway along the northern coast of Crete, on a low hill about four stadia across. It might have been a coincidence; but, then again, who knew?

In 1939, when Marinatos first published his theory that widespread destruction on Crete and the subsequent eclipse of Minoan civilization might be connected with the volcanic destruction of Thera, the editors of the journal *Antiquity* attached a disclaimer to his report, pointing out that, in their opinion, "the main thesis requires additional support from excavation."

Clearly, the volcano had exploded, but in 1939 there was no reliable evidence pointing to when; 1,000,000 B.C. seemed just as valid a guess as 1600 B.C. There were, of course, stories of buildings and pottery under Theran ash; but almost all the evidence had been destroyed by mining operations, and what remained was largely dismissed as tombs built into the ash by civilizations that had dwelt on top of it, rather than preexisting buildings entombed beneath it when the volcano awoke. If Marinatos could find undeniable proof of Minoan streets and houses under the pumice, he'd be able to fit the eruption into a Bronze Age time slot, and a major argument against his theory (critics claimed that the eruption had occurred thousands of years before Minoan civilization even existed) would fall away.

His plans to carry out the excavations called for by the *Antiquity* editors were thwarted by would-be empire builders in Germany,

Italy and Japan. Amid the commotion, on September 18, 1940, one of Mussolini's bombers attacked a freighter as it steamed past the volcanic island in the center of Thera Lagoon. Bound for Athens, the ship carried nothing more strategic than pumice from the Fira quarry, and no armaments of any kind. The Italians dropped most of their bombs anyway, missing the freighter by a quarter mile and striking the volcanic cone instead. Awakened by man's violence, nature responded with violence of her own. Rocks and thick black dust laced with steam exploded out of the ground in an uprushing column that towered and expanded hideously, drawing a curtain of smoke over the freighter. The crew of the bomber concluded that they must have targeted a munitions ship, reported their success to base, then turned north and winged home. Roman radio bulletins reported the victory. The citizens of Fira laughed and offered a toast in the direction of the cone. They'd seen the ship emerge from under the smoke screen and leave Thera Lagoon unharmed.

As the war raged on, the cone continued to cough dust and boulders into the air daily. The sea around it steamed and swirled, one afternoon milky white or yellow, the next morning scarlet, or green, or violet. By the time World War II ended, the Greek civil war was beginning to heat up, so it was not until the 1960s that Marinatos could return to the Cyclades.

He stopped first at the island of Kea, one hundred miles northwest of Thera, where he had an opportunity to examine a Minoan house firsthand. Marinatos hoped it was merely a foretaste of what lay beneath Thera.

For five years Professor John Caskey, assisted mostly by graduate students from the University of Cincinnati, had been excavating a building that seemed to have been stamped flat around 1500 B.C. (give or take 150 years) and later rebuilt by Greek inhabitants. The people who came after the Minoan destruction had put down a slate floor over the mud-filled wreck and added new stones on top of its half-buried foundation. When Caskey dug below the floor, he discovered that some of the older, deeper stones were jostled out of line with the upper part of the building. The Greeks must have reset the exposed upper portions of the walls before building on top of them. Two feet below the new floor lay a mixture of broken pottery, fresco pieces, mud and hundreds of household objects belonging to the original owners. Both Caskey and Marinatos agreed that much of the debris had fallen from upper stories into the basement; but Caskey attributed the collapse to a great earthquake, while Marinatos saw evidence of a tidal wave. The site was only a short walk from the sea, and life-sized terracotta statues of thin-waisted ladies had been pounded and scattered throughout the entire length and width of the building. Many of the fragments were heaped up in a corner, as if they had been shifted about by a large mass of receding water. Interestingly, the later, Greek, inhabitants must have uncovered the head of a terracotta lady as they laid down the slate floor, for they appeared to have displayed it as a valued work of prehistoric art until their new house was burned to the ground.

Marinatos took note of an odd convergence of Greek literary tradition and archaeology. At the request of the people of Kea, the poet Pindar (fifth century B.C.) wrote the *Paean* in honor of King Delos, which begins: "I tremble at the heavy-sounding war between Zeus and Poseidon. Once, with thunderbolt and trident they sent a land [Thera?] and a fighting force [Minoan naval power?] down to the Tartarus [a sunless abyss below Hades]."

During the summer of 1965, as Arthur C. Clarke and Wernher von Braun traveled overland to Monolithos, a Theran beach notable for its black volcanic sand and the slab of rock that towers over it, Spyridon Marinatos was exploring the entire island, trying to figure out where he should dig first. His attention was settling on the southern coast of Thera, because the vanished mountain would have shielded it from the Etesian winds, which blew from the northwest from May through October, and were, in autumn, violent enough to push people over cliffs and prevent ships from sailing. Marinatos had no way of knowing this, but the zigzag and cluster arrangement of streets and buildings in Minoan Thera—a pattern of construction that produces baffles against wind—had been reproduced exactly, and independently, across thirty-six hundred years. Function dictated form. The buried city resembled twentieth-century Fira more closely than the ruins of Minoan Crete or Kea.

If he were going to build a Minoan palace, or anything else on the island, Marinatos decided, he'd start at or near the village of Akrotiri, which, even in 1965, was known to be a safe harbor for ships in a storm. Just as important, any homes there would have received the sun all winter long, whereas buildings on the north shore—certainly not choice real estate—lay always in the shadow of the mountain.

On the south shore, in Akrotiri, Marinatos noticed circular stone troughs that the villagers were using as drinking basins for their donkeys. (They are still there today.) He recognized them as ancient stone mortars carved out of local volcanic rock. Each weighed more than a hundred pounds and could not have been dragged very far from wherever it was that they had been discovered.

A villager named Arvanitis showed him a field where plowing had been halted by a pile of rectangular stones. The field overlooked a ravine where, each year, for dozens of centuries, the winter rains had sent a flood running over the buried city. The dusty stream bed was now a scratch in the earth's surface. It cut right through the building excavators would one day know variously as "the grainery" and "the bakeshop." On the hill overlooking it, the pile of blocks that obstructed plowing was in fact the eroded second-story window over the entrance to the building Marinatos would, with a typically scientific lack of enterprise, name "West House" (because it was located at the westernmost edge of his first survey).

And then Arvanitis led him to a place where he'd seen the ground collapse under the weight of a passing donkey. Not very far away were rectilinear depressions where the soil had subsided long ago. Marinatos began to picture underground cavities, buried rooms in the upper stories of houses whose roofs, from time to time, were still caving in. He drew a breath, held it reflectively, and let out a laugh. On this very spot, some poor donkey had plunged through thirty-six hundred years of earth history—straight into somebody's bedroom.

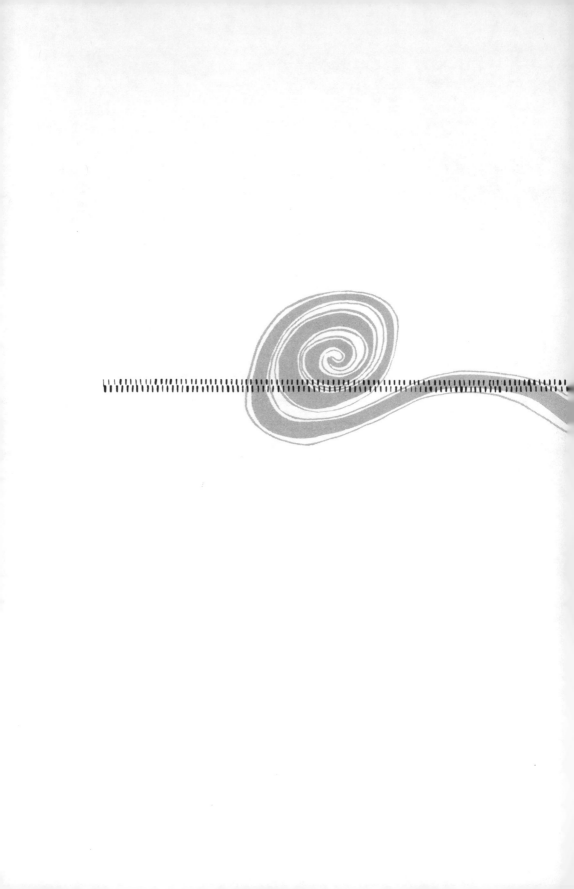

RESURRECTION

I suppose most excavators would confess to a feeling of awe—embarrassment almost—when they break into a chamber closed so many centuries ago. For the moment, time as a factor in human life has lost its meaning. Three thousand, four thousand years maybe, have passed and gone since human feet last trod the floor on which you stand, and yet, as you note the signs of recent life around you—the finger-mark upon the freshly painted surface, the blackened lamp, the farewell garland dropped upon the threshold—you feel it might have been but yesterday. Time is annihilated by little intimate details such as these, and you feel an intruder.

—Howard Carter, *The Tomb of Tutankhamen*

DESCENT
INTO
DARKNESS

Paleontologists proudly admit it, while archaeologists, for some obscure reason, tend to deny it: we are the biggest storytellers in the world, we who stroll down the corridors of time.

The chief differences between the two groups are the time scales we are accustomed to and the amount of preserved material available to us.

In New Zealand I once found a cliff face that exposed a two-million-year column of Miocene time, during which the earth's climate changed, changed utterly, and the last of Glen Affric's giant crabs changed with it before descending into extinction. Through intervals of time as small as fifty thousand years, I could follow the story of a diminishing subtropical fauna and its replacement by creatures more closely resembling some of today's species. It was one of the richest sites I had ever encountered. One level contained nautilus shells. Another was studded with crabs turned wholly into stone. In another, dating back twelve million years, I found the twelve-foot-long skeleton of a whale. It looked familiar, yet somehow different.

When I travel back further, ninety-five million years to Cretaceous time, I am happy to find just a scrap of dinosaur bone, or a thumbnail-sized nugget of amber.

But even scraps can sometimes tell big stories. In 1977, paleontologist Gerard R. Case brought me to a field in Sayreville, New Jersey, where he had discovered the ruins of a ninety-five-million-year-old forest, out of which, in only a few short months, a housing development would begin to rise. A world had slept, unnoticed, within the earth, and from this world, before it was lost forever beneath someone's backyard, Case plucked a random snapshot of the Age of Dinosaurs.

It was a quiet death. Ninety-five million years ago, tree sap flowed gently over a bee. A wing twisted, and the insect struck a disquieting pose that would last almost forever.

Sealed in a transparent coffin of amber, the bee remained intact while the tree that had become her tomb grew old, fell and decayed. The forest, too, disappeared. The Atlantic Ocean doubled its width, the general temperature of the universe fell 0.3°F and, unbelievably, a little insect came through unchanged in appearance—flash frozen in amber—with every facet of her eyes, every vein in her wings, every spur on her legs perfectly preserved. Her tomb was less than three quarters of an inch across, yet it had proved to be twenty-four thousand times more permanent than any pharaoh's.

The story told by a thumbnail-sized piece of amber goes beyond death in a Cretaceous forest. All previous fossil evidence placed the origin of flowering plants about sixty-five million years ago, around the time of the last dinosaurs.* The bees came later, and

* Some theorists have even proposed that flowering plants killed off the dinosaurs. According to this theory, dinosaurs were unable to taste the bitter, poisonous alkaloids in new plant species, and are said to have gone belching and farting into oblivion. One of the major flaws in this theory is that it does not explain how seagoing saurians went extinct eating those same plants, or how so many organisms unrelated to dinosaurs (including many of the world's oyster

the oldest previously known fossil bee had been found on the Baltic coast and dated back only forty million years. The New Jersey bee was more than twice as old, and pollen-collecting spurs on its hind legs told us that bees were already highly adapted to flowering plants, and they to bees. The coevolution of plants and bees therefore went back at least to the middle Cretaceous, over one hundred million years ago—meaning that we must now re-write the story of when flowers first appeared on earth.

To a paleontologist, the preservation of a single bee is a spec-tacularly beautiful event. It allows me to sketch honeysuckle and hyacinth into a landscape where previously there were only tree ferns and evergreens. I look at another fossil, a fragment of tooth, and I see ostrich dinosaurs no taller than me darting among wild roses. Two pieces of rock, two interconnected stories.

To a man who walks among the dinosaurs, Thera and ancient Egypt happened a very short time ago. Yet I have met an Egyp-tologist who has difficulty dealing with time scales beyond thirty thousand years, and seems to think that the Atlantic Ocean opened up only about sixty thousand years ago. In terms of time, pa-leontologists and archaeologists think in two different languages; but in another way we are the same. It is what I (something of a paleontological chauvinist) prefer to call paleontological think-ing: the ability to see a story in a piece of rock or a chip of carbonized wood, to look at another rock and find another story, and to find still another in a splash of pigment.

Once you begin thinking like a paleontologist or an archaeol-ogist, stories jump out at you from every direction. It becomes habitual. A single insect has the power to recolor parts of the big picture; but the painting over of old ideas is no cause for despair. If you are wise, and pay attention, the picture is always in motion, always being revised. New truth is very noble. The explorer Rob-ert Ballard once explained to me that he viewed science as a game

species) also disappeared. The final nail in the theory's coffin, of course, would be evidence that dinosaurs and flowering plants had coexisted for tens of millions of years before the extinctions.

of Clue, where the butler did it in the library with a candlestick. And the game was, could you figure out the truth with a minimal amount of information?

That was autumn, 1985. We were at sea, looking at hydro-thermal vents and poring over the first photographs ever taken of the Royal Mail Steamer *Titanic*. It was probably the world's most peculiar introduction to marine archaeology. It was a whole new world to me, emerging as I was from a background in Cretaceous and Miocene time. Rejoining pieces of the *Titanic* from photographs of a mile-long debris field, then retracing the pieces—whole again—back to the horrible spasm of her last three minutes, turned out to be the easiest paleontological puzzle I'd ever solved. I had been used to reading stories from only a handful of preserved scraps. Here were literally tens of thousands of clues and they were all laid out for me in pictures. Nothing even had to be dug out of the rocks.

To me, Thera is much like the *Titanic*. Because Thera is so young, the earth has not had time to erase the record. A single room contains hundreds of artifacts, hundreds of little stories. They overwhelm. Yet unlike the *Titanic*, Thera has never moved me to tears. Here I know only the excitement of exploration, of trying to solve a puzzle. Enough time has passed to obliterate all names, all sorrows, all personal horrors. There are no mountains of court transcripts, no newspaper accounts of what happened here, no survivors' diaries—and no survivors for me to look up. There is nothing except nameless, mostly faceless people, the city they left behind and legends, misty and vague and strange—which leads me to one of the most interesting (and downright enjoyable) aspects of archaeology and paleontology: you never really know if the story you have constructed from the rocks is correct. There is always an element of doubt. A story can change on you at any moment. We may learn, for example, that the world really began when a turtle climbed on top of an elephant's back and said, "Let there be galaxies."

Again, this would be no cause for despair. It would just make the universe a little more fascinating than we'd previously believed. To begin with, we'd want to know what was underneath the elephant.

The ancient, traditional stories of the world's beginnings are based upon faith and are immutable, even in the light of evidence telling new stories. The paleontologist's view of the world is based on doubt and must leave open even the possibility that new evidence may eventually prove one of the traditional stories to be true. It is detective work beyond the imaginings of any detective; and the pursuit, every step of the way, is both exhausting and exhilarating.

SUMMER, A.D. 1967

Nanno Marinatos was seventeen years old when she went with her father to the island for the very first time. It was a beautiful place, there was no question about that. A thousand feet below Fira, a white stone tossed into the lagoon was still plainly visible thirty feet down. It fell twenty feet more before it faded to turquoise, then violet, and then was gone. The drop-off at the edge of the lagoon was as steep as the cliff above, and Nanno guessed that the stone she cast in must have fallen at least as far.

On donkeyback she followed the same snaking path von Braun and Clarke had ridden two years before. A villager had told her that the beasts were inhabited by souls sent from purgatory. They paid penance for their sins by toiling up and down that thousand-foot cliff, as often as twenty times a day. Throughout Greece and the Cyclades, donkey owners warned stubborn or rebellious animals to behave, "or I'll send you to Thera."

All trips between Fira and Akrotiri had to be made on foot or donkeyback across fields, for no roads connected the two villages as yet. Nanno's lasting impression was of freshly chiseled cliffs, and of an island that was still wild, virtually untouched by the twentieth century. The hotels, restaurants and nude beaches beneath

Mesa Vouno were nearly two decades in the future. Tourists only rarely came to Thera. There were no nightclubs, swimming pools, taxi cabs or street lighting. Nothing except occasional white-walled houses, little blue and white churches, the nine-year-old ruins of two thousand earthquake-ravaged homes, and miles of wilderness.

Nanno was more impressed by nature, by the unspoiled beauty of the island, than by the possibility of finding a lost civilization. She had been living with her father at excavations since the year she was born. Her earliest recollections were of crawling around in tombs. To make life more interesting for her, after he'd found and documented a new pocket of bones and artifacts, Spyridon Marinatos would palm an object—a ring or amber pendant fashioned by craftsmen of the sixteenth century B.C.—and bury it where she could find it. She remembered being about five years old and running after her father, proudly proclaiming a new "discovery," and how good it made her feel to be contributing to his search for the past.

At seventeen, archaeology held very few surprises for her. It was still fun, and she did enjoy camping out in the deserts, with no bright city lights and only the cold stars burning overhead; but after seventeen years, one hole in the ground was beginning to look much like any other. It had not occurred to her that this was a whole city they were about to excavate. It was a dawning realization that, for the moment, eluded even her father. He did not guess, as he tunneled into a ravine, that at least three hundred years of excavating lay ahead. But the city was patient; it could wait.

There were sprawls of dead vegetation out there, grasses mostly, with an occasional cluster of vineyards and fig trees on well-watered ground. The volcanic soil was rich in mineral nutrients and the island would surely have been a paradise if only it received more water. Instead, it was a desert. Rainwater percolated far too fast through the coarse volcanic rubble, as if through a sieve. Only rarely did it form dendritic streams like those that had, for thousands of years, run down through Akrotiri and carried away more than a hundred feet of volcanic tephra, doing most of the digging for Spyridon Marinatos long before he was born.

In summer, Thera was completely waterless, save for winter rains stored in underground cisterns and the weekly arrival of a supply ship at Fira. Life was not easy for donkeys and figs and grapes: they were obliged to survive on used bath water.

The elder Marinatos's expanding network of ditches and tunnels revealed a jumble of carefully hewn stones, some of which had once backed smoothly plastered walls. Amid the cracked plaster were traces of frescoes lying in thousands of pieces below the ground. Delicately painted ceramic vases turned up in the ash. Holes pierced their rims, holes through which long-vanished lengths of knotted leather must have been strung. The owner of this house was wealthy enough to own hanging gardens. It must therefore have been more than a house. Marinatos began to believe that he'd been extraordinarily clever, or lucky, or both; managing to locate a Minoan palace on the first try.

The spade struck a startlingly modern set of kitchen utensils, including drinking cups and frying pans. They were green now, having oxidized in the groundwater and congealed as a single body of copper and tin deformity. Oxygen, mold and bacteria had consumed Minoan Thera, removing all organic furnishings. Hanging plants, cedar beams, screens of wooden doors that slid in bronze tracks, pinewood toys and all the life of the city were reduced to cold, black streaks of carbon. Water and scavenging microbes had penetrated the soil, fanning out, searching, and tapping dark stains that used to be rattan beds, tripod tables and perhaps even people. The spade probed deeper, exposing to daylight a tiny cave, a cavity in the earth where, a very long time ago, cedar or some other organic substance had vanished. Spyridon Marinatos had heard of such things in the ruins of Pompeii. He poured plaster into the hole, let it harden, then scratched away the surrounding ash. What remained was a fossil impression of an ornately carved wooden table.

A bronze cup was found inverted, and near it a stone lamp with traces of lampblack still visible. There was a finger smudge upon the blackened surface, undisturbed for thirty-six hundred years.

A wall had pitched forward and crushed ceramic ewers, jars and a marble chalice. On top of the wall, and on top of everything else, lay a thick blanket of volcanic ash. But beneath the fallen stones, amid the broken housewares, Marinatos could not find a trace of ash or pumice. Now, in a three stage argument, he could tell a story about the last days.

Two facts he knew: ash fell over the whole ruin; ash did not fall under the collapsed wall. This led to a third fact he did not know before: the place was in ruins before the ash fell. An earthquake had destroyed the building; thereafter a snowstorm of pumice-ash buried everything. Even though he hadn't been there to see it, it must have happened that way. Logically it had to.

"I remember my father taking me around the site and telling me stories," says Nanno Marinatos. "There were rows of man-sized pottery jars emerging from the white ash, still standing upright in exactly the positions they had been left some thirty-six centuries ago. Chemical analyses would soon show some of the vessels to contain the dried remnants of wine and oil. Others were filled with fossilized grain, and another with the shells of land snails.

" 'A Bronze Age feast!' my father observed. He then noted that the shells belonged to a species found today on Crete but not on Thera. And he began to ask: 'Were the snails imported from Crete, or did Thera once have its own population of these creatures, and were they all exterminated by the eruption?'

"He was a great storyteller. He described to me how he imagined things were, how people lived. Those are some of my most cherished and moving memories of him: evenings on the terrace, overlooking the sea. We lived at the excavation, we built a little house and we lived there. And I'd sit with him under the stars, asking questions and hearing stories. Sometimes it was the very absence of things that told part of a story. No traces of bodies were turning up. And while finely crafted tools were buried in the ash, there was no hint that gold or silver or anything of real value had been left behind, as in the case of the Roman towns buried by Vesuvius in A.D. 79. My father began to suspect that the Therans had not left in a hurry. They had time to take their valuables with them before the ash came. And I think I am correct in saying that we had taken the first steps toward an infinite diverging regress. We felt as if we knew nothing. We knew more about the dinosaurs, and of them we had only scraps to read from. There, on Thera, we had evidence piling up all around us—whole buildings filled with answers, and yet every new story produced questions, leading to questions, leading to more questions. We were in fact learning many things about Thera; but every answer seemed only to double the size of the original problem. The buildings were preserved. How? What happened to the people? They left, apparently. Why did they leave? An earthquake had knocked down their walls, it seemed. *When* did they leave? And the questions went on.

"The stories my father told me—they weren't stories in the sense of a narrative about people walking through dead corridors; they were told more in the tone of a search. They were questions put together in such a way as to make the problem exciting. You wanted to dig the next day and see—was he right, or must we revise the theory?

"One particular night of storytelling I will never forget. He was looking out across the Aegean and reciting Homer. Not that the *Odyssey* had very much to do with the catastrophe at Thera, but it's just that he was conjuring up the past and making it come alive. He was able to recite quite big chunks of the *Iliad*, the *Odyssey*, and Apollonius of Rhode *Argonautica*—

And then straightaway, as they moved swiftly over the great Cretan deep, night terrified them, the night which they call "the pall of darkness." No stars nor moonbeams pierced the deadly darkness. It was black chaos coming down from the sky, or some other darkness rising from the inmost recesses of the Earth. They did not in the least know whether they were voyaging on the water or in Hades. Helplessly they entrusted their safe return to the sea, to carry them whither it would.

"And he'd make the atmosphere very evocative of the life of these disappeared people.

"During those first months on the island, we did believe we had stumbled upon a buried palace, like the one at Knossos. The huge storage jars were so similar in size and number to the ones excavators had found at the Cretan capital that my father had an idea that Thera was going to be more of the same. By the middle of the second digging season, however, in the summer of 1968, this picture changed. Boy, how it changed! He started digging random try-out trenches, and everywhere he dug, he would find more storage jars, more pottery and more walls. And we began to realize that it was in fact a city.

"From that moment my father was on a high that had no equal. It was the most exciting excavation of his life—perhaps the most exciting excavation of any archaeologist's life since Howard Carter entered the tomb of Tutankhamen. And the excitement went into getting people, calling together lots of collaborators. For restoration work the most skilled masons, fresco and marble workers, metallurgists and carpenters were needed. He sought out physicists with backgrounds in carbon dating, and chemists who could determine the precise ingredients in a vessel of dehydrated perfume or wine. And above all, he required additional grants from the Greek government to finance the surge. There began a hurricane of hyperactivity, and the focus of the storm was on the south shore of Thera, and the eye of the storm was Spyridon Marinatos."

And you may ask yourself: How did I get here?

—David Byrne, "Once in a Lifetime"

APPRENTICE

The road to scientific endeavor is not always a straight and narrow line. Only rarely does a child decide at age nine that she wants to be a zoologist, or a paleontologist, or an astrophysicist. Sometimes it's a purely serendipitous outcome.

"I must tell you, it was quite accidental," recalls Doumas. His father was a landless farmer who cultivated other people's property, "so I didn't have any cultural, any literate background in my family." When Doumas went to Athens to study, he started with theology, which did not interest him. In the summers, he would usually travel home to Patras by ship. "And in one of those journeys, during the year I was studying theology, and I was not so happy about it . . . I met a Danish student . . . and we started talking. And when he asked me what I was studying, I said theology, and—he didn't understand—he said, 'Archeology?' And then I thought—'Ah, that's what I want.' "

SUMMER, A.D. 1955

If you asked Doumas to look back across three and a half decades and describe the single most exciting archaeological moment of his career, he'd surprise you, because it had nothing to do with uncovering Atlantis. It happened one summer when he was still a student. He was assisting the archaeologists of Athens in excavating a plot of land that was going to be developed.

The archaeologists had found a cistern from the fifth century B.C., and Doumas was participating in the clearing away, an inch at a time, of dried mud from the buried chamber. Near the bottom his group descended through brownish-black ooze into a layer of red sediment. This was the original, old Athenian layer, and since Athenians seemed to have been in the habit of littering their cisterns, it was hoped that ancient rubbish would allow the excavators, via cross-referencing with the contents of other cisterns, to more accurately date the site.

Doumas was combing through dirt that had been thrown out by workers when he came across a black shard of pottery, no more than two inches on any side. When he rubbed it with his fingers he saw letters. He spat on it, rubbed harder and read:

$$\mathsf{K L E \, S \quad \Xi A N \Theta I \Pi \Pi O}$$

—Kles Xanthippos

"Kles" was the ending of the name Pericles . . . Pericles, the son of Xanthippos—it was the great Pericles, who had been Athens' leader during the height of its glory, but whose actions had led to the Peloponnesian War, which destroyed Athenian democracy.

According to historical traditions passed down from ancient Greece, the little black chip was an *ostrakon*, which was used to cast a vote against someone, and from which we get the verb "to ostracize." Someone had voted against Pericles, but history teaches us that he was not ostracized, because the vast majority of the votes were cast in his favor. The great Pericles of Athens—too clever and charming for his own good, or for anyone else's.

How had the ostrakon come to be cast into the cistern? Had the man who hoped to blackball Pericles changed his mind on the way to the ballot box? Perhaps it was merely discarded at some point after the vote. There was no way of knowing. But still Doumas could, to some degree, live through the atmosphere of elections in classical Greece when somebody—a fanatic Pericles-hater, perhaps—angrily scratched the name into a triangular black shard.

AUTUMN, A.D. 1968

The technological landscape of 1968 was at least as alien to A.D. 1989 as 1989 would be to 2010. No one really knew what the surfaces of the moon, or Mars, or Europa looked like. Today, visions of other worlds are commonplace. In 1968 such places were simply far away in outer space, much as Keftiu, Thera and the Pillars of Hercules must have seemed very far away in the west to ancient Egyptians.

Far away in the east, American military forces, equipped with jet fighters, attack helicopters and batteries of computer-assisted missiles, were being defeated in Vietnam by generals who lived in caves and consulted astrologers. At home, on the shores of the Mediterranean, no one owned a personal computer, and the thought of assembling a private videotape library containing all of one's favorite movies belonged to the realm of science fiction. Color television was the newest innovation, and only now was it becoming widely accessible. Cable TV did not exist as yet, and the closest things to a music video were Disney's *Fantasia* and Kubrick's *2001: A Space Odyssey*. Computer games, Sony Walkmen, automatic tellers, FAX machines and private satellite dishes were nowhere to be seen. News tended to travel more slowly in those days. Secrets were easier to keep.

Doumas had been in England when the first shovelful of dirt was cleared from the buried city. He was nearing the end of a two-year sabbatical leave, tying up the last loose ends of his doctoral dissertation for the University of London. As an undergraduate at the University of Athens, Doumas had studied with Spyridon Marinatos. Had there not been a continent between them, Doumas might have noticed that his former professor seemed to be very excited about something—that is, if one could track him down. He seemed to be moving in four different directions simultaneously—one day to Turkey, the next day to Athens, or the Cyclades, or Iceland.

Then, in October 1968, Marinatos dangled an irresistible carrot in front of Doumas's nose. He asked him to come to Thera and assist him. He'd been digging for two seasons, and in every inch things had been turning up.

Things . . . wonderful things.

Nothing could be seen from the surface, because Marinatos had dug long trial trenches—twenty feet deep in places—and then filled them in again to protect his finds from the weather. He had no desire to repeat the mistakes of the excavators at Pompeii.

Pompeii had survived amazingly intact for nearly two thousand years beneath the ash of Mount Vesuvius. While all the other Roman cities, unrepaired and uninhabited during the long nights of barbarism and the Crusades, dropped apart stone by stone on the earth above, graffiti remained legible on Pompeii's bathroom walls. In the nearby town of Herculaneum, a case of fine glassware sat half unpacked, and the sick boy in the shop of the gem-cutter lay in his elegantly veneered bed. His mother had prepared chicken to tempt his appetite, and placed it on the table beside him. Less than six feet away she'd set up a chair and a loom, so she could look after him while she worked. The room was silent and dark, the boy's mother missing, his lunch untouched.

It was a disquieting fact, but a fact nonetheless, that the city's exposure to open air and sunlight and rain had inflicted more damage than the volcano that originally snuffed out its life. The polluted air of Naples brought down acid rain, which assailed marble facades and limestone statuary, dissolved them, softened their features. Frescoes faded, flashburned by the sun 365 times yearly. In a patrician's courtyard, dark green foliage had begun a crazy garden. A savannalike growth crept up the avenues and climbed the steps. And whatever nature had failed to destroy, the visitors were making fast work of. Several pieces of gold jewelry had disappeared. John and Martha from Omaha had scratched their names over ancient Roman graffiti. The Department of Antiquities had begun to loan genuine bronze furniture to theatrical companies, including the one that would soon be filming *I Claudius*, and, in what would eventually become one of history's great marvels of unconcern, was neglecting to ask for the return of priceless artifacts. A table with bronze goddesses for legs, ancient charcoal braziers and tripod oil lamps were dispersing quietly in the general direction of America, and by 1980 at least one Hollywood home would be furnished almost exclusively from Pompeii.*

No, there would be no Pompeiis here, Marinatos had decided. So he filled in his trial trenches, all except one. As long as the city lay in state in its white shroud, the earth would protect it from looters and winter rains.

The single, open trial trench was actually a guarded tunnel into the side of a ravine. Roofed over by more than twenty feet of volcanic matrix, it resembled an old prospector's mine, complete with ceiling supports assembled hastily from cracked wooden beams. The corridor snaked along an ancient street, intercepting walls and huge vases filled with remnants of barley, tuna, roasted snails, sesame seeds and buns.

On one side of the passage, the front of a house had been preserved; it reached up to a height of at least two stories. A

---

* At least some of the furnishings were moved to the San Francisco Bay area, just in time for the 1989 earthquake and fire.

doorway appeared to have been thrown propitiously open (significantly, there were no indications that the Therans slept with locks on their doors). The entrance was a solid wall of pumice, deposited in layers of fine ash interspersed with pebbles. What lay a few inches beyond that opening was anybody's guess; more vases certainly, perhaps even the skeleton of a Theran splayed out in the awkward attitude of sudden death.

It would be a long time, however, before the archaeologists would be able to see for themselves. Marinatos's original idea had been to honeycomb the entire city with corridors. He felt that it would have made quite a dramatic public display, being able to stroll down the streets of a city beneath the earth. But in the tunnel, dried white pumice was crumbling and falling. And the island was still volcanically active. The thought of being in, or on top of, or near this tunnel the next time Thera grumbled was unnerving.

It was impossible to enter the houses through their doors for fear of undermining the rooms above. Even if the upper floors did not crash down upon them, the archaeologists would be unable to dig into the rooms in such a way as to preserve everything they discovered, or to document the exact position of every find. They could easily destroy far more than they learned.

Far, far more, Marinatos understood. Even if he dug down very carefully from the fields above, his exploration would be like reading backward in time, backward through the pages of a book, except that he'd be forced to destroy each page as he read it. But at least by descending from above, he'd be able to photograph and write down very carefully everything he saw before he disturbed it beyond recall. The archaeologist had to make a precise copy of the book as he destroyed it.

The tunnel had seemed a good idea at the time. Beautiful. Very romantic. But he'd have to abandon the concept. He decided to excavate the city like an open pit mine, but with one important difference: this mine would be roofed over as he stripped away the earth. He was going to protect the buildings from wind, and rain and weeds—and especially from that two-legged species of weed that loots.

A doorway lay before Marinatos, and beyond it a room full of mysteries. All he had to do was dig through to the other side. But he must be patient. More than twenty years later, that same doorway would still be filled with pumice, even though the entire building would then be sticking out of the earth, sheltered beneath a vast canopy of tin and steel.

Marinatos must have felt a curious kinship with Howard Carter, who had crouched in another tunnel, in front of another doorway—blocked, plastered and sealed—on another autumn afternoon forty-six years before. Six inches away, on the other side, lay the tomb of Tutankhamen. But archaeologists are an extraordinarily self-possessed breed. Without hesitation, Carter filled in the tunnel, posted his most trusted guards in front of it and sent a cable to his mentor in England:

AT LAST HAVE MADE WONDERFUL DISCOVERY IN VALLEY. A MAGNIFICENT TOMB WITH SEALS INTACT. RE-COVERED SAME FOR YOUR ARRIVAL. CONGRATULATIONS.

Then he waited two weeks for Lord Carnarvon to arrive.

Like the entrance to the Theran house, the burial chamber in the Valley of Kings had been long forgotten. And they were contemporaries, or very close to it: Tutankhamen was probably buried little more than 150 years after Thera. But beyond this there was no comparison. Visitors to the burial chamber of Tutankhamen were surprised by nothing so much as its smallness.*

---

* Relative to Ramses VI, whose tomb lay on top of Tutankhamen's and completely obscured it, Tutankhamen was a short-lived, minor god-king, whose obscurity had preserved his burial chamber from looters—which was just as well for us, because his golden death mask radiates a depth of feeling that is lacking on portraits decorating the outer golden sarcophagus, and in almost every other work of art created by man. If the earth were to be destroyed tomorrow and only fifty works of art could be saved, this would surely be one of them. It would have company from Thera.

It could easily have been hidden in a little corner of the city Marinatos and Doumas were about to excavate. The contents of Tutankhamen's tomb told the story of only one child-king's death, with glimpses of his life; the rooms that lay beyond the pumice walls of Marinatos's tunnel contained thousands of stories waiting to be told.

Throughout the 1968 digging season, the progress of the excavation ran parallel with construction of the tin roof. The rate at which the archaeologists penetrated the city depended on the nature and age of the layers they were digging through. The youngest layer was chalky ash interspersed with fist-sized splinters of black and red bedrock. The final explosion had deposited the mixture on top of the city's protective layer of white pumice in a matter of seconds. As this uppermost layer contained no archaeological evidence, it could be removed quickly—with a bulldozer—at the rate of at least six feet a day.

When they reached the shroud of pure pumice they had to proceed more carefully, because it was there that the ruins began. The buildings had once been supported by timbers, but during the three dozen centuries that had passed, the wood had oxidized into black powdery nothingness, leaving only hollow fossil impressions in the volcanic debris. Where buildings were preserved to heights of two and three stories, Marinatos's people had to seek out the negative impressions of wall reinforcements and door and window frames as they worked their way cautiously from the roofs to the basements. The negatives were easy to find. They were literally everywhere, empty spaces where the wood once was. If the excavators were lucky, when they broke open a void it would run through layers of densely packed volcanic dust. Then they could pour concrete down the hole, where, reinforced with a skeleton of steel wires, it assumed the shape of the missing wood, adhered to the ancient masonry and, as the surrounding dust was removed, supported the walls as the wood once did. If they were unlucky, the phantom wood ran through layers of coarse pumice stone, and the concrete leaked out into spaces between the nuggets, then hardened into an amorphous blob.

Most of the time they were lucky. Some of the ghostly voids turned out to be impressions of rattan beds, carved wooden tables, door frames, the backs of chairs with their wood grain intact, and pieces of looms with their clay weights still in place.

Inside the buildings, progress down into the earth was very slow. Ash had piled up on the houses like enormous drifts of snow, caving in their roofs. The archaeologists were often forced to make their descent through collapsed floors and fallen debris, a hopeless mixture of voided wooden beams, colored stone tiles, pottery, frying pans and frescoes broken into thousands of pieces.

The frescoes had not fallen alone. Their wreckage was mingled with pottery and metal objects, and the excavators had to proceed painstakingly, for they would never find the fragments of plaster lying conveniently face down on the floor. Instead, the bits and pieces—large and small—were mixed in with the fallen walls, lying sometimes flat and sometimes obliquely between bricks, other times standing on edge in the ash. Because they had to study how the pieces had fallen, Doumas explains, the archaeologists had to "move slowly, and to dig mainly with a brush."

"I remember, I spent my first winter in the storeroom, playing with the fresco bits, like a jigsaw puzzle." It was becoming clear that shattered pictures—which could tell much about the people who lived in the city—were accumulating faster than the archaeologists could restore them. A team of seven full-time experts did nothing but reassemble frescoes. Twenty years later, they'd still be there.

The digs continued at a slug's pace. The archaeologists were descending through parts of West House sometimes at the rate of one inch through a single square yard of ash during a whole day, because every fresco piece, before it was removed, had to be consolidated, drawn and photographed so that one could easily find pieces to rejoin in the laboratory. There emerged a rigid, unbreakable equation: the more carefully they proceeded with the dig, the more quickly the restoration progressed in the lab. Pictures of ancient fleets began spreading across tables in the back rooms; ladies wearing gowns and gold necklaces; two adolescent boys with boxing gloves—the first known record of gloves of any kind.

Black, red and white were the dominant colors, just as Plato had said they would be. The decorations were everywhere; not a single home was without them. One wall had shown antelopes at play, beautifully stylized, yet at the same time so true to life that they could be identified as a species found only in East Africa. Did Minoan sailors venture so far?

Such questions were turning up daily as, with paintbrush and toothbrush, the men descended into Thera inch by tedious inch.

SUMMER, A.D. 1988

Only two buildings have been fully explored. Ten homes now protrude from the ash beneath the tin canopy, and at least seven others remain buried here. To a nonarchaeologist or a nonpaleontologist standing here for the first time, Minoan Thera would not look particularly impressive. The five buildings open to the public occupy barely more than an acre. And unlike Knossos and the temples of ancient Egypt, these have not been reconstructed for tourism. The archaeologists have chosen to excavate and display the buildings in their true historic context: that is, as an earthquake-ravaged settlement abandoned by its citizens shortly before the final upheaval. A home owner returning to the excavation after nearly four thousand years would enter the ruins of West House to find the place much as he left it: the same storage jar in the same position, precisely where he—perhaps as an afterthought—had righted it before walking out the door for the very last time; the familiar, man-sized *pithoi*, urns, near Delta

House on Telchines Road still standing in rows; and a fallen wall resting on smashed vases, exactly as he'd last seen it.

To appreciate the meaning of this single acre of buildings, you have to send your imagination out under the miles of surrounding pumice. There is no sign that the outskirts of the city have been discovered. Almost a mile away in the west, miners at the Mavromatis quarry have found almost identical, as yet unexplored homes. Other ruins have been found a quarter mile in the opposite direction, and may also have been part of the city. This one exposed acre represents only the tiniest fraction of what actually exists out there: buried treasure beyond anything you can imagine, thousands of undiscovered frescoes, perhaps even a library with its works intact on clay tablets. A cemetery would make possible the first scientific study of Minoans as people: their height, their weight, their eating habits, work habits and medical history—even exact reconstructions of once-living faces from skulls. Teeth and bones are storehouses of biological secrets. To a paleontologist, it is through skeletons that the dead truly reveal their lives.

WINTER, A.D. 1968

Doumas had known that Spyridon Marinatos was a strict and difficult person to work with. But when you went to an oral exam after Marinatos had ravaged your thesis, after he'd made you go back to the library and the laboratory and do it over and over again, you knew that other critics would have no surprises for you. If it passed Marinatos's criticism, it could stand up to anything, and that was one reason the best of students gravitated toward him: he was tough, but you learned—boy, did you learn!

The first thing Doumas learned at Thera was that Marinatos was a very nice person to sit with over salad and Coca-Cola discussing what the Russians and the Americans might do on the moon next year, but as a boss he was even more demanding than he'd been as a teacher. It was like walking through a minefield: wrong moves lay in every direction.

For a start—just for a start—their political points of view were not only different, they were fundamentally opposed. Greece and the Cyclades were under the dictatorship of Georgios Papadopoulos. Marinatos supported the dictatorship. Doumas was against it, but he did not discuss—dared not discuss—his views openly. Marinatos did not appear to be the type who would throw a young man to the wolves, but these were politically dangerous times, when people were known to disappear in the night.

So Doumas's relationship with his boss was not close. He sensed from the beginning that Marinatos wanted to be at the center of everything, and the professor was sending some very strong signals that he did not like an apprentice showing too much scholarly interest in the site. When Doumas saw the wealth of artifacts emerging from the earth, he announced that he'd seen similar things before, in Crete and elsewhere, and suggested that he revisit some of the other islands and see what other Minoan artifacts people had been digging up, so he could do a better job of recording and analyzing the newer, Theran material.

Marinatos did not reply.

Two days later, he made Doumas a different proposal. As head of the Department of Antiquities, he would provide some money for Doumas to conduct a minor excavation at another site on the island—to keep himself busy, as Marinatos put it.

"No, thank you," Doumas said. "I don't want to dig *anywhere* because I have plenty of material from our previous work and I'm not fond of excavations before we cope with what is being excavated. I think the material that is produced by this site is such that I will never cope with it and an army like me will be needed."

An army like him!

Marinatos must have recognized from the start that young Christos Doumas was much like him. They shared an irritating habit of looking at the same pottery vases or scraps of broken rock that others had been looking at for a lifetime, and then connecting them in startling new ways to come up with answers different from everyone else's. They also shared a certain degree of arrogance, and a keenly developed flexibility of mind that enabled them to organize knowledge, not merely to store it as any machine could do. They had a capacity for divergent thinking that made them almost an alien species. Marinatos knew from experience that Doumas's training would be difficult. Many teachers responded with an immediate, almost instinctive revulsion toward the aliens in their flock and sometimes set out to destroy them. A few—very few—tried to nurture them.

But by keeping Doumas close, was Marinatos not, at the same time, taking a terrible risk that his former student might compete with him and eventually seize control of the excavation? Like it or not, both men possessed superior abilities and, like it or not, superior abilities had a tendency to breed superior ambition.

There had been many victories during Doumas's first season at Thera. He was embarking on an adventure beyond all reasonable expectation; and, yes, an army like him would be needed to cope with the buried city. But the thirty-two-year-old apprentice was a disappointed, worried man as 1968 drew to a close.

He was disappointed because he realized that Marinatos did not trust him and was trying to restrict his scientific activities, acting perhaps out of fear that a former student might steal his glory. And he was worried that one wrong move would get him banished from the excavation that was now yielding up the first view anyone had ever had of the most romantic, fascinating ancient world of all, the lost civilization which, if Marinatos was correct, had been the basis for the legend of Atlantis.

As word of their finds spread, and journalists began to converge on the island, Doumas realized that he'd best keep to the shadows. Whenever TV cameras moved toward him, he referred the reporters to Marinatos. When Marinatos was absent, he acted only as spokesman, requesting that they not mention his name at all. And bit by bit, Marinatos came to trust him.

Doumas also noticed that the professor did not graciously accept the fact that anyone else had ideas. Thus, whenever he wanted to propose a new way of organizing the descent into Thera, he made a point of saying, "As you said, Professor, we have to do this. . . ." The tactic worked every time.

Doumas had come to Thera with finely tuned instincts for scientific endeavor. But that was not enough. Without fully realizing what he was doing, Marinatos was giving him a thorough education in the politics of science. Many scientists had come before Doumas, and many would come after him, without ever gaining such knowledge. As a result, careers would end before they began, secrets would remain undiscovered, books unwritten; and the world would always be a little less knowable, and perhaps a little less beautiful.

The great aim of archaeology is to restore
the warmth and truth of life to dead objects.

—Philippe Diole, *Herculaneum*

A
ROOM
FULL
OF
LILIES

SUMMER, A.D. 1969

Arvanitis, the man who had led Marinatos to the upper stories of West House, owned the land several hundred yards west of the excavation. One morning, while digging a well, he broke through to a layer of salty groundwater and beach sand beneath Thera's pumice shroud. Marinatos became very excited when Arvanitis showed him the sand. He had concluded by this time that the city beneath Akrotiri village and its surrounding farmlands must have been one of the Minoans' wealthiest, most vital ports. If the lessons of Greek, Egyptian and Roman excavations were any indication, only captains, merchants and kings could afford to have their homes painted by professional artists. Some of the Theran homes had stood four stories tall, and there were pieces of frescoes in all of them.

"Oh, yes; a wealthy port," Marinatos recorded in his notebooks. "No doubt about it."

But there was, as yet, no indication of where the harbor and ships might have existed. The citizens of Akrotiri farmed now on land that had once been empty space. Near Arvanitis's wells, grapes grew in the dizzying heights a hundred feet above ancient Thera's tallest buildings. South of the excavation, nineteen-year-old Nanno Marinatos often strolled along beaches where, in Minoan times, there had been open sea. The pumice and rubble that covered the city had totally transformed the coastline.

"I see what the salty sand means," Spyridon Marinatos told Arvanitis. "Your well is an eighth of a mile inland, and the prehistoric coastline came in at least as far. Today you grow vineyards between the ridge of Mavrorachidi and the foothills of Mesa Vouno. About thirty-six centuries ago, somewhere on (or rather, under) your farm stood loading docks and ships. Sheltered by Mavrorachidi Ridge in the west, by Mesa Vouno in the east, and by the vanished mountain in the north, it would have been the ideal location for a harbor."

Marinatos could not know it at the time, but a wall painting on the ground floor of West House depicted a fleet of Minoan ships in a harbor surrounded by hills. One of the hills looked hauntingly familiar. It was Mavrorachidi Ridge, seen in profile from a vantage point near Arvanitis's well. The painting also showed a watchtower or lighthouse perched on the ridge crest, in precisely the spot where the foundations of a Minoan tower lay in pieces.

Though Marinatos had seen much during his first two years of digging into Thera, he had imagined even more. He liked to believe that the fossil impressions of one or two abandoned ships lay beside the still undiscovered docks. In many ways, the search for the lost Minoan fleet was the realization of a childhood dream. He'd never outgrown his love of ships, or his curiosity about those days when the seas were new, and large portions of the earth remained unexplored. Whenever he sailed between the Cyclades, he liked to imagine himself in the company of Odysseus and Orpheus, retracing some small part of Homer's Mediterranean

Odyssey; but like most of the things that were important to him, he tended to keep it to himself.

During the first years on Thera, living facilities at the excavation were barely more than an improvised night camp in the wilderness: three tents, one shack, collapsible chairs and tables, latrines and a portable power generator. Excavation was concentrated in the summer months, when graduate students were free to assist.

One August afternoon, Doumas unfolded a canvas lounge chair and lay down in the shade of Marinatos's shack. He was hot and tired. It was 3:00 P.M., the height of the day's heat, time for the customary Theran siesta.

"Look! Mr. Doumas!"

One of the guards was pointing excitedly out to sea. Doumas sat up and looked.

"See, Mr. Doumas? A huge boat is passing near the coast."

"All right. There is nothing strange about it."

"I am going to tell the professor."

"No, don't," Doumas warned. "He will be furious. He is relaxing now, and very easily he gets irritated."

"No, no. He likes to watch such things."

Now, that was something to think about. It was hard to believe that Marinatos would invite an intrusion upon his siesta merely to look at a ship. Doumas was amazed to hear it. "So, go," he told the guard. "Go. I want to see this."

Doumas settled back in his chair, pretending to be half asleep, watching from a safe distance.

"Professor, come out and see a big, huge boat!" the guard called.

To Doumas's surprise, the professor appeared at the door with his binoculars. "It is a French design!" Marinatos called out. He knew just what sort of machinery the ship must be transporting, and he began to describe it at length.

He's like a child, eh? thought Doumas.

The apprentice looked on with fascination. Marinatos was full of surprises. Doumas had known for a long time that the professor loved to watch birds and study their habits. And for endless nights he'd stood under the stars with him, listening to dissertations about ancient mythology and the constellations, and galactic empires, or some such craziness. And now this unexpected knowledge about (almost obsession with) ships. Was there anything the professor did not have a hand in? Anything at all?

Across the street from West House, in the building known as Delta House, the archaeologists were slowly stripping away sheets of volcanic ash. The Delta House excavation would be particularly memorable. Its rooms were filled with layer upon layer of household objects: containers, fossil chairs, tables, beds, the remains of yet another hanging garden and a carbonized wicker basket embedded in a cocoon of tephra. There were fish hooks, an awl and surprisingly familiar domestic utensils, including frying pans, cooking pots, cups, saucers and knives that, apart from being made from bronze, would not have earned a second glance in a present-day kitchen. Delta House was also memorable because it was about to yield up a room whose walls were undamaged frescoes of Theran spring before the great eruption—the work of a forgotten genius. Instead of using solid borders and subdivisions to break up wall paintings, as at Pompeii and in most modern estates, the artist had shaped the room into a unified landscape, in which red lilies grew amid volcanic boulders, nodding as if bent by a gentle breeze, while swallows darted above them.

One other thing was going to make the emergence of Delta House memorable. By the time the building was completely excavated, Spyridon Marinatos would be dead, and his final resting place would be within room Delta 16.

The room was full of powder and dust. Whenever they encountered pockets of fine-grained ash, they knew beautiful things lay below, as the ash had unusual preservative qualities. The top of a wall was still covered with uncracked plaster painted red. And because they had encountered chips of red plaster in every room, (leading Doumas to question whether the entire wall had been smoothly plastered and painted red from floor to ceiling), they had no idea that the ash concealed anything so beautiful as the Fresco of the Lilies. Two feet lower the spade revealed a stone slab shelf. Packed pumice ash still preserved an image of the vanished wood that once held it in place.

"The part of the wall above the shelf was covered with this red color, and underneath the shelf was the fresco," recalls Doumas. "As we removed with a brush and knife, layer after layer, thin traces of dust, flowers started appearing—leaves, then the head of a bird. That was very fascinating to see, gradually, the whole thing."

Near the floor a void pumped full of liquid plaster at high pressure hardened into a perfect cast of a rattan bed. Even the leather thongs that hooked the fur and hide mattress to its wooden frame had left distinct fossil impressions. The bed was only five feet six inches long, confirming Marinatos's suspicion that the Minoans, like most Mediterranean people of ancient times, were tiny.

Unless it was a child's bed—

Early one morning, when the room was completely cleared of ash and artifacts, Marinatos entered. For a long time he stood alone on the cool floor, looking at the first sunshine slanting through the skylights, feeling the stagnant air. Overnight, a million droplets of moisture had gathered on the plaster, threatening harm. By noon the dew would be melted off, and the air would be dusty and close.

"The fresco wants to express the spring season," Marinatos would later write.

Amorous twittering swallows were depicted in naturalistic poses, leading Marinatos to believe that the artist must have been very familiar with their habits.

But swallows no longer live on the island, he noted, although there had been ample time for them to return. There were swallows on all the nearby islands. But not on Thera. Perhaps the eruption was so terrifying as to uproot even their migrating habits.

A race memory, deep wired and instinctive: Stay away from Thera!

The fresco gave an impression of infinite charm and love of nature. It was, perhaps, a religious allegory with origins reaching across the Mediterranean to Egypt.

Nature worship? Marinatos wondered.

The idyllic scene was reminiscent of that he and Nanno Marinatos used to see in Egypt. In the tombs, often there were frescoes of lotus and papyrus. The Egyptian religious landscape was a vast expanse of marsh. The afterlife was viewed as a field of rushes. If the Minoans were nature worshipers—in fact, that would be ironic, considering what nature eventually did to them.

Doumas was not satisfied with the emerging theories about Minoan religion. Many archaeologists were too quick to attach religious significance to the remnants of ancient cultures. A room full of lilies became a religious landscape, and the room itself a place of worship. A fresco of ships on the sea became a ceremony. Some religious influence occasionally crept into even modern paintings; but wasn't it possible that the wealthy Minoans were just decorating their houses?

One had to wonder how archaeologists thirty-six hundred years in the future would interpret the tacky ceramic skulls that were becoming all the rage on the mainland. They were used as ash trays. It was a hoot at death and cancer; but it was easy to imagine some university brat attaching a death cult or some such nonsense to the skulls. He would not likely guess the truth: that it was just an expression of morbid European humor.

What fascinated Doumas were the stylistic irregularities in many of the frescoes. He detected the hand of more than one artist. There were even hints that the errors of one hand had been corrected by another.

Apprentice and teacher? Doumas wondered.

As more fresco bits came together to form pictures, an interesting pattern was emerging. There must have been whole schools of artists, Doumas would later write, and they must have had a high degree of freedom in the execution of their works, if not in their choice of themes—which were undoubtedly selected by those commissioning the work. As a result, the paintings are remarkably unconventional, full of vitality and the breath of inspiration. This is in marked contrast to the frescoes of Crete and Egypt, which adhere to the rigid conventions imposed by the ruling palaces. On Thera, artists did not appear to have been monopolized by the monarch, but were patronized by the more affluent members of a competitive society, who paid them to enhance their surroundings, and perhaps also to impress the Joneses next door. Competition would thus have been encouraged among the artists, each of whom strove to create something innovative and original.

And the plants on the wall were, if anything, original—though in a way that Marinatos and Doumas could never have anticipated. They were hybridizations of the white madonna lily and the red lily. Genetics 101, Marinatos observed with astonishment. Could such knowledge already have existed?

The rock landscape upon which the lilies grew represented an unhoped-for opportunity for geologists: red and black lava flows solidified against white. This was Thera before the big eruption.

Marinatos had been talking with the geologists. Most of the eruptions before that last day seemed to have been relatively gentle lava flows, as are commonly seen on the Hawaiian islands today. The last major flow went back to Neanderthal times. Thera had been heavily forested then, but the lava probably burned up most of the trees on one side of the island. There were, as yet, no bones to prove that the Neanderthals or anyone else had been on Thera to witness the destruction.

"It's fascinating to explore deep time," Marinatos thought aloud. "But seen against the time scales of the earth, we are not the focus of attention. We're on our own, eh? We are just invisibly small specks set against a universe whose beginning is nowhere and whose center is everywhere."

"So there," he said to no one in particular. "Now who do you think you are?"

The wall paintings were a great challenge—and achievement—for the restorers. The archaeologists did not want to leave them in place for the next Theran earthquake, so they decided to remove the walls intact to the museum at Athens. Of course, observed one excavator, if civilization is ever stupid enough to fight a nuclear war, all the great works will go, along with the cities: the Room of the Lilies in Athens, Tutankhamen's golden death mask in Cairo, all gone in an instant—a vapor in the sky. Perhaps they would be safer buried where they were found, or scattered throughout the world in private collections. After all, if so many great works had not been gathered together in the library at Alexandria, they would not have been so easily destroyed by a depraved archbishop.

"We archaeologists were very worried and scared about the fate of this—to see three walls removed altogether," Doumas recounts. "But our restorers had quite long experience working on Byzantine frescoes, and tried to calm us down." Still, Marinatos made all of the restorers sign an agreement spelling out the procedure in detail before he let them touch anything. In the end, everything went smoothly. The whole thing was cased in plaster and plastic reinforced with asbestos gauze, then gently lifted out on ropes. If you go to Athens now, you can see the room, minus its furniture, reconstructed exactly as it was found.

The excitement of the descent into the Room of the Lilies must have been unbearable. "We didn't know the composition of the painting from the beginning," Doumas recounts. "So, you find a leaf, and you say, 'Oh, that's good!' You don't know what is coming and you are always in a sort of excitement and an agony—what is hidden?" At one point, the excavators were surrounded by hollows where the bed had been, and two hundred and fifty pieces of pottery. Crouched between urns, they had to watch the head of a bird rising from the ash without moving.

But Doumas remembers a time when Marinatos himself exploded. It was in the north entrance of the building where the fresco of the lilies was found. The excavators had uncovered a group of pots that were lying there, and they had halted work to make a record of what they had found. "All of a sudden, I hear Marinatos shouting, 'Come and see! You are all blind? You didn't see anything?!' On one of the vases there was an inscription which we hadn't seen, incised. It couldn't be seen unless the light fell obliquely, and Marinatos happened to pass at the right moment. And that was the only inscribed vessel we had found so far—it is exhibited in Athens, too—of Linear A script, the script of the Minoans. And I remember his excitement about this, because it was the first unique text we had."

Linguists have long noted a peculiar fact: that the ancient Greek name for Thera (Kalliste) and the names of such famous Greek places as Olympus and Knossos are not truly Greek. Some scholars, including J. V. Luce, detect a Mideastern flavor in the names. There is something distinctly Luvian about them, suggesting that people who once inhabited the hills of Turkey spread out of Catal Huyuk into the eastern Mediterranean, bringing their language and their custom of bull worship with them to Crete and Thera.

On the eastern tip of Crete, directly in line with a thick blanket of ashfall that covers the seafloor and gets progressively thicker as one moves closer to Thera, stands the wreckage of Zakros. Unlike the people of Thera, the inhabitants of Zakros left behind their best bronze tools, even their gold and silver. A palace fell in smoke and flames, and if we are, as some archaeologists have suggested, to attribute the destruction to plunder by invading peoples, we must ask why the plunderers failed to do what plunderers do best: plunder.

When explorers found the settlement in 1962, the wealth of the palace was still scattered across the floors of demolished, rubble-filled and forest-covered rooms.

The artifacts of Zakros suggest prosperity and a wide-ranging sea trade. There are elephant tusks from Syria, carved pink granite from Upper Egypt, copper ingots from Cyprus.

Throughout Crete are clay tablets written in Linear A. A ceramic disk was found in the ruins of Phaistos impressed with 241 pictogram images of heads in profile, with Mohawk haircuts, sheepskins, hammers, sickles, birds, flowers, walking figures and buildings. The list of glyphs goes on. What is amazing about this disk is that the figures—representing phrases and sentences whose meanings we do not know—were not written on its surface, as by a scribe; they were actually stamped on it, with the same stamp being used repeatedly, as needed. The Minoans of 1650 B.C. have therefore provided us with the earliest known example of something akin to movable type. They were literate people; but no one knows what it is they were trying to say. We can dig into their homes and learn how they lived. We can even see pictures of daily life in their frescoes, but unless Thera holds a library with many points of reference surviving on clay disks and tablets (heaven help us if they wrote on paper scrolls, for water and oxygen and bacteria would have consumed them), the Cretan tablets will remain mute on the triumphs and tragedies of those last great days.

During the Linear A period, when Thera's buried city still gleamed in the sun, art themes tended to focus on women in long dresses, on flowers, rivers and leaping dolphins. Then the ash streamed out of Thera and passed over the Aegean, falling across eastern Crete and accumulating three inches deep. Agriculture must have ceased there, ceased utterly, and the archaeological record shows a subsequent population shift toward the relatively ash-free western third of Crete. There also seem to have been migrations upland, away from the sea. New towns sprang up (we know this because the lowest, oldest levels of their garbage dumps are filled with cracked pieces of newer, post-Thera pottery styles), and the western settlement of Kydonia swelled from the influx, swelled from a town to a city.

During the next century, the writing, too, shifted. Linear A

was replaced by the Linear B style, the so-called Minoan II style—
which the British scholar Michael Ventris has shown not to be
Minoan at all, but a primitive form of classical Greek. The translations that inevitably followed this realization have proved to be
one of archaeology's great disappointments. It was, at one time,
hoped that rooms full of clay tablets unearthed in post-eruption
Crete were parts of libraries that had survived; but there was no
trace of a late Minoan *Macbeth* among the piles of Linear B tablets.
Instead, the only texts to survive the ravages of time are the
etchings of clerks: records of harvests, inventories of sheep in
given flocks, useless statistics—"the contents of a few wastebas-
kets in five Cretan capitals," laments one archaeologist.

But this much the tablets tell: the newer posteruption Linear
B script is found both on Crete and mainland Greece. The pre-
eruption Linear A is found only on Crete, Thera and some of the
other Cyclades. The traditional archaeological consensus held that
Linear A evolved into Linear B and was given to the Greeks by
the Minoans. Quite the opposite may be true, because on Crete,
the shift from Linear A to Linear B appears to have occurred
abruptly (in a century or less, following the eruption) and the
two scripts are totally alien to each other: one cannot translate
Linear A using Linear B as a guide. The next obvious theory is
that the Minoans, having grown weak, were subject to occupation
by mainland Greeks, who imposed their language upon them—
and their artistic styles, too. Bulls were no longer shown being
captured with staves and nooses; instead, they were felled with
spears and arrows. Lions, too, were felled in arenas that evolved
slowly into colosseums, and sometimes men were felled there as
well.

The Minoans left few traces of violence in their art. The Greeks
(and later the Romans) adopted it as a major theme. The oral
history of the early Greeks was that of a people forever locked in
chaos, devastation and expeditions of conquest—a history that
comes down to us through the writings of Homer, Plato and
Herodotus. Had Thera never exploded, the Minoans might have

been able to keep the Greeks at bay indefinitely. "But for more than two centuries they [the Greeks] had been gathering strength," wrote Marinatos. "They extended their influence into the Aegean world and, while it was expedient to do so, lived in apparent harmony with the Minoans, drawing heavily upon the Cretans' skills to forge their own civilization. Finally, when flames, ash and tsunamis came bursting up from the sea, and Cretan power began to diminish, the opportunity for complete Aegean control was open, and they moved swiftly into Knossos."

Without preserved libraries, we have only the substance of archaeology from which to read the story of Thera and Crete. Plato's *Timaeus* and *Critias*, along with a dozen or more biblical references to cities going up like the smoke of a furnace and being swallowed by waves, may be records of eyewitness accounts, but from very far away, and the accounts were blurred by many centuries of telling and retelling before finally being committed to paper.

The Greek historian Herodotus recorded the oral history of the Praisians, who lived in eastern Crete until 140 B.C., called themselves "genuine Cretans," spoke a non-Greek language (Minoan?), and recounted a tale that augments the implications of the Linear A to Linear B transition. According to the Praisians, as told by Herodotus, Crete was depopulated, and then "other people settled, and especially Greeks." Before the depopulation, King Minos is said to have set out in search of Daedalus, trying to extradite him from Sicily to Crete.* Evidently, Minoan influence extended as far west as Italy, but King Minos was not very popular there. The Sicilians killed him, inviting a naval assault that seems to have failed. At some point, a Minoan fleet was "flung ashore" on the southeast coast of Italy. The survivors mysteriously gave up hope of ever returning to Crete. Instead

* There were several Minoan kings named Minos, spread over many centuries (just as the descendants of early Hebrew tribes have been passing down the names Abraham and Isaac for thousands of years), and Greek history seems to have blurred them together as a single King Minos.

they established their own community, becoming "mainlanders instead of islanders." They built the city of Hyria, and one has to wonder why.

If the castaways could build a city, then surely they possessed the energy and resources to build at least one boat and send a small party of sailors home to Crete, where they could summon a rescue expedition. Such an attempt must have been made (though Herodotus does not record it), and it is possible that the real reason the Minoans decided to stay in Italy was that when their boat returned to Italy, alone, it brought news that Thera had sunk, that Crete and other nearby islands had not fared much better, and the Minoan navy was, for all practical purposes, no more. Seen in this light, the storm that flung the builders of Hyria ashore might very well have been a Theran tsunami.

"It is said that Minos went in search of Daedalus to Siciania, which is now called Sicily and met his death through violence," wrote Herodotus.

Some time later at the instigation of the god, all the Cretans except the Polichnitans and Praisians came with a great armada to Siciania and besieged the city of Kamikos for five years—a city now inhabited by the Akragantines. In the end they were not able to capture it, and shortage of food compelled them to disperse, so they departed for home. On their voyage they had got as far as Iapygia [the southeast coast, or "heel," of Italy] when a great storm struck them and flung them ashore. Their ships were smashed to pieces, and since there seemed no available means for returning to Crete they remained where they were and founded the city of Hyria. They changed from Cretans to Messapians of Iapygia, and from islanders to mainlanders. In Crete itself, bereft of its inhabitants, as the Praisians say, other people settled, and especially the Greeks.

SUMMER, A.D. 1988

No close-up, firsthand accounts of those last days were written; at least none that survived. We do not know the name of a single person who lived here. But there are beautiful wall-paintings in what must have been living rooms and bedrooms, judging from the furniture. Nameless women, some of them in long, flowing dresses, others with their breasts bared walk among antelope, swallows, papyrus bushes and lilies. None of these things live any longer on the desert above. A pond or a stream can no longer exist up there. Yet the frescoes show us what once was. A canal or river cuts through a landscape of palm trees. A town with a harbor lies at the foot of a mountainous region, and from the mountains a stream emerges, encircling the town. It is this picture of the stream that encourages Doumas to continue his search for the mysterious water source that seems to have filled the city's cisterns and flushed constantly through its sewage system.

In an upper story of West House, toilets and a ceramic bathtub have been discovered. The wooden toilet seat has vanished, of course, but a cylindrical clay pipe is still intact. It drained directly down into the sewers beneath the street, making several bends along the way. That means water had to flush through it, and above the concrete seat frame there is a recess in the wall that looks for all the world like it could have held a water tank.

The invention of the flush toilet has always been credited to Sir Thomas Crapper in the eighteenth century A.D. I wonder what he'd think today if he knew the phrases that carry on his name. In any case, the Minoans seem to have preceded Sir Thomas by about thirty-four hundred years.

As for West House's water source, the most fascinating speculation I've had so far is that steam pressure from volcanic vents was used to drive some kind of pumping apparatus that lifted water to rooftop cisterns on Theran houses, or that the steam itself could have been piped up and condensed in the cisterns.

Some of the homes are riddled with pipes. Perhaps, on the way to condensation tanks, where the steam eventually became bath water and toilet water, it was routed, during the winter months, through wall pipes, which radiated heat into the rooms. Of course, this would have required valving the steam up through external pipes during the summer months. We'll have to look for traces of those pipes, assuming they ever existed. (A word of caution: like most hot speculations, this one is probably wrong).

Who knows? The truth must lie buried somewhere in this city. My most rational guess is that we will find remnants of elevated aqueducts.

One thing that interests me is that the frescoes on the walls of West House show a city—this city—surrounded by a stream. Plato, in his description of Atlantis, described such a city, and said that it was blessed with streams of both hot and cold water (*Critias*, 113e). There is nothing conclusive here, or particularly compelling; perhaps nothing more than an interesting coincidence.

The frescoes also show a fleet of ships sailing from Thera to another harbor city, perhaps to Libya or Syria—or even Egypt, where the Great Pyramid at Giza was still gleaming white, gold and silver. In the painting it looks like a fleet of trading ships. It has all the signs of having been a peaceful contact, not a voyage of conquest.

Though we may never be able to read the Linear A script of the Minoans, their buried houses and housewares, and especially their frescoes, teach us much about them. They were master builders, even by the standards of Egyptian pharaohs. Across the water, at Knossos, some of the apartments appear to have been equipped with showers, and beneath the city ran water conduits so large that a man could stand in them without stooping. On imperial Crete alone, a quarter of a million people must have lived, as many as forty thousand of them in Knossos, where the palace warehouses contained row after row of man-sized storage jars with a total capacity exceeding 315 cubic yards.

For all their prosperity and power, the Minoan kings and merchants appear to have behaved very differently from their Egyptian neighbors to the south. There is no evidence that anyone in the buried city, or in any of the other Aegean cities, erected huge statues or tombs to celebrate his own glory. Labor, natural resources and architectural talents were spent on the population at large, on the construction of warehouses, ships, irrigation systems, country villas and fine-quality homes, rather than on pyramids, gold-lined tombs and Theban monuments serving an elite minority of kings. Egyptian frescoes and limestone reliefs boast official pomp and pageantry, and expeditions of conquest against the pharaoh's enemies. The frescoes of Thera and Crete do not even hint at the achievements of Knossos' ruling class. The Minoans evidently placed greater importance on idyllic scenes of people at leisure, or carrying fish to the market, or engaged in boxing, bull fighting and other sports. Almost half of the rooms are decorated with scenes of flowers and wildlife.

Even in the treatment of their slaves, the Minoans seem to have been unusual. There are no traces of crude shacks and slave-row barracks in or around Knossos, Zakros, Phaistos or Thera. Writing more than a thousand years after the Thera upheaval, the Greek philosopher Aristotle, referring to historical texts still extant in his time but now lost, noted that Cretan slaves enjoyed the same rights as other Minoan citizens, except the right to bear arms.

Judging from the frescoes, Minoan women, too, enjoyed far, far greater social freedoms than their counterparts elsewhere in the world, who were often treated as property—slightly more valued than slaves and cattle—and were traditionally beaten to death for even minor offenses. By contrast, a fresco at Knossos depicts women and men together at a public festival, extending hands to each other and engaged in animated conversation. The women are elegantly dressed in fine lace, jewels, beads and ribbons. Some of them wear narrow-waisted skirts, others what look curiously like bell-bottom pants. In another fresco, women participate alongside men in sporting events. Both hold the reins of chariots. We may never know for certain if women were treated as true equals, with such privileges as the right to vote (assuming Minoan citizens ever voted on anything); but in the frescoes, at least, they seem anything but oppressed.

It is plain that the Egyptians respected the Minoans as a refined and civilized people. All other foreigners, especially those from Israel, Syria and Babylon, were despised by Egypt as "barbarians who are an abomination to God." Only the seafarers from Crete and the Cyclades were honored in Egyptian tombs with a separate, respectable name—the *keftiu*.

It's anyone's guess what the keftiu might have become, if only they'd been living somewhere else, anywhere else. They were the newest and brightest people around, and they disappeared in a geological nanosecond under a volcanic death cloud, during an interval of time astonishingly small, even by human standards. Yet the very thing that destroyed them has preserved them, has allowed them, after more than thirty-six hundred years, to speak clearly. Their apartment houses, their beds, their storefronts and paintings showing us who they were and how they dressed—even their sewage systems—have lived almost forever.

Quickly slipping on a dress, seizing anything that came to hand, I put on a long fur coat and rushed out into the lounge, wearing only slippers on my feet. But before doing so, I did a most extraordinary thing, when I regard it in calmer moments: I took everything I had in the room in the way of jewelry and dresses, and threw them into my trunks, shutting the trunks and locking them, and closing my stateroom windows and shutters. . . . The *Titanic* seemed the biggest thing in the world. All was calm and still, the reflection of the lights on the water, passengers leaning over the rails, life-boats lowering, strains of music filling the air: Nothing to indicate the coming horror.

—from the memoirs of *Titanic* survivor Edith Russell

THE
EDITH RUSSELL
SYNDROME

AUTUMN, A.D. 1988

Gold and silver and the finest bronze tools were apparently removed from Thera—as if the island was abandoned before the final upheaval. The complete lack of skeletal remains in the ash—not even the bones of a single cow or donkey—suggests to most archaeologists that everything of value, including the animals, was loaded onto boats and taken away. To date, the only silver found within the excavated area consists of two thin rings. A fragment of gold leaf appears to have been lost and forgotten on a jeweler's floor. It is difficult to say what lies beyond the excavated area, except that a gold-inlaid dagger unearthed about a half mile away hints to me of more chaotic events in distant parts of the city. Yet here the overall impression is of order, even though before the exodus there were earthquakes powerful enough to crack stone staircases and bring down several buildings. The people might have left in haste; but in several rooms tables and pottery jars seem to have been righted before they—before they what? Drew the curtains and closed the doors?

As they sifted through their earthquake-ravaged homes, salvaging what they could, a kind of nesting behavior seems to have taken over; deep rooted, strange and totally instinctive. They probably were not even aware of what they were doing.

Nanno Marinatos tells me that when the man-sized storage jars were first discovered near Delta House, on Telchines Road, all of them were standing upright and were neatly arranged in rows, in exactly the positions they had been left some thirty-six centuries ago, before snowdrifts of ash entombed them. Surely a quake powerful enough to topple walls only a few dozen feet away would have overturned the tall pottery jars as well. If there was anything that could be used from the jars—and it is notable that, when found in 1967, they were still full of food—all the ancient Therans needed to do was spill out what they needed and move on. Instead, they wasted time. They salvaged inefficiently, illogically. There was no need to tidy up, to right fallen objects and arrange them in rows—but that seems to be precisely what they did, before leaving the place forever.

I am reminded of a passage I read in Edith Russell's memoirs. Looking back, she was astonished by her own behavior on that incredible night when the *Titanic* went down. She made sure a friend's puppy was petted and kissed and lying comfortably under a blanket before locking it in a bedroom on the sinking ship. She tidied up her stateroom, putting her clothes neatly into their trunks, closing her windows and shutters, and making certain that everything was in order before retreating to the safety of the lifeboats. I wonder if this is a normal human reaction to leaving a doomed ship or city: to make a last-minute, hurried effort to tidy up and leave the place presentable. I wonder if anyone was delayed by this instinct just long enough to miss the last boats casting off from Thera. Yet we've seen no traces of bodies, so far, no evidence that anyone was left behind and buried with the city, at least that portion of it lying under the tin roof.

Nesting behavior . . . it's powerful stuff. I remember an unusual ebb tide in Mexico, near the ruins of what seems to have been a Mayan seaside resort in Cancún. Outcrops of coral normally located four feet below the lowest tide were climbing into the air. On one of them, a butterfly fish lay dying in the sun. The instinct to stay with its little familiar patch of "home" had overridden every impulse to follow the retreating waters to safety. I am reminded again of Edith Russell, who went to her room, started dusting the furniture, and decided, at one point, not to abandon the warm and familiar *Titanic* for the uncertainties of a night adrift in a lifeboat, even as the coming catastrophe became obvious.

In Iceland, when a new volcano sprouted near Helgafel in 1973, sending a wall of lava advancing slowly into the town of Heimaey, escapees were willing to sacrifice precious minutes—hours even— to sweep ash from their floors, to fold towels and blankets and store them neatly away in closets. Rescuers recall carrying people away from their homes kicking and screaming, insisting that they be allowed to sweep just a few more square feet of ash from some corner of a doomed kitchen, so strong was the need to—to what? To leave a good impression? To leave the place presentable to a lava flow?

I have seen this phenomenon in Niagara County, New York, where the Occidental Petroleum Corporation buried 192,800 tons of chemical wastes—a volume sufficient to completely fill the *RMS Titanic* and still have enough left over to sink the aircraft carrier *Nimitz.** Occidental sold part of the land to home builders. The homes are still there, looking now like something out of *Salem's Lot.*

The place is called Love Canal, and when I stepped onto 101 Street last June, the school athletic field was a savannalike growth

---

* The wastes consisted mostly of phosphorus-based explosives, banned insecticides (including C-56, Mirex and Kepone), and such banned defoliants as Agent Orange. There are 1,171 sites just like Niagara County scattered throughout the United States, and probably twenty times as many throughout the world. Let us all hope that Niagara's Love Canal region is not a foretaste of much of our planet for the next century.

of grass and scattered bushes—one turned out to be a badly de-
formed maple tree whose leaves were Siamese twins. Scrub grass
had grown wild and tall in the front yards of two-story houses,
then spilled knee-deep onto Muller Court. Crickets sang in it, and
I found myself searching instinctively for a recently cleaned window, a reshingled roof, a footpath through the weeds . . . but I
could see nothing that gave any indication of living human beings.

These people had fled not from a volcano, but from poisons of
their civilization's own creation. In almost every home on 101
Street there had been cancer, birth defects, rare blood disorders,
respiratory problems, kidney or liver disease. The cause of the
abandonment was different, but the survivors of Love Canal behaved much like the Icelanders at Heimaey, much as Edith Russell
had behaved on the *Titanic* and as the Minoans apparently behaved
at Thera.

What makes the absence of people at Love Canal particularly
eerie is that the homes appear to have been cleaned before being
abandoned. Though most of the rooms have been emptied of their
furnishings, and a decade's worth of dust has accumulated, the
wooden floors were definitely given a last, fresh coat of polish.
The houses were made to look presentable, as if for prospective
buyers.

In one house, couches, clothing and TV sets have been left
behind (the owners probably believed them to be permeated with
poison). Dishes are stacked in cupboards. Towels, sheets and pillow cases are carefully folded and stored in closets. Everything is
in order, except for the rain-drenched carpet under the broken
front windows, an overturned garbage pail on the porch and vines
growing over the pail. In the backyard an above-ground swimming
pool has collapsed. The wooden pool deck, too, is caving in, and
the fence separating this yard from the one next door has begun
to rot. On the bottom of the pool lie about six inches of stagnant
water, into which black, oily matter seems to be percolating from
somewhere far below. The water smells like dry-cleaning fluid.

In front of the house, facing oncoming traffic that isn't coming anymore, stands a rusting yellow sign: SLOW DEAF CHILD. In the next block, there is another. Then another. Then somewhere else another.

What story does that tell?

And what did they do before they drew the shades, locked the doors and drove away? Edith Russell syndrome: they dusted their houses and put their towels in the closets.

"It is not uncommon," says historian Walter Lord, "for people caught in the worst of times to remember to take care of absurd details. You see it all the time. The woman who pauses to look in the rearview mirror and brush her hair after a minor traffic accident is a common example.*

"What is the purpose of this response? Does it serve any useful biological function? Or is it just one of those quirky, harmless little instincts that we seem to have inherited from our fabulous furry forebears?"

I've given a good deal of thought to that, and yes, I do believe the syndrome has come down through time as a useful piece of genetic luggage. One can imagine many ways of handling poten- tially terrifying events. Panic is one of them, and that way lies the greatest danger. When coping with a crisis, the safest thing to do is to keep your head, and the brain seems to be programmed to help us do precisely that. We can mentally shield ourselves from even the most bizarre catastrophes by latching onto some- thing relatively normal.

Edith Russell, for example, was fully aware that the *Titanic* was doomed, yet her nesting urge probably prevented her from fleeing blindly in some random direction, perhaps to the nearest over- crowded lifeboat, or even into the water. Only a handful of twenty-two hundred people who shared the slanting decks with

---

* Walter Lord prefers to call this behavior "*Yorktown* syndrome," after the U.S. aircraft carrier sunk at the Battle of Midway. Several sailors have told him that, as they prepared to climb down the ropes and abandon ship, hundreds of them removed their shoes and arranged them neatly in rows, as if expecting to return. In the backs of their minds, even as they did this, they were aware that they were doing something very strange.

her panicked that night. There were not, as one might imagine, swarms of human bodies jostling each other in a rush to the lifeboats. Indeed, many of the boats were lowered half empty, and Edith noted that the people standing by on the deck could not have been more orderly if gathered in a church.

Having switched suddenly, unconsciously, from a mode of wanting to flee in any direction, to a mode of taking care of "home," gave Edith breathing space. She recalled years later that as she drew the shutters and dusted a table she was acutely aware of, though no longer overwhelmed by, her surroundings. The world seemed no longer frightening or bizarre. Panic was banished.

Edith was not alone in her strange response.

"Now, there's something you don't see every day," a man remarked, as the liner settled nose first into the ocean and a smokestack crashed down barely two yards from him. Elsewhere on the ship, men in evening clothes were still playing cards on tables that slanted twenty degrees, a bartender stayed by his post, announcing that drinks were now on the house, the band played on and an immigrant girl on her first sea voyage returned to her room for a pillow, so she could take a nap on the lifeboat. She'd convinced herself that everything was perfectly normal, that ships operated somewhat like trains: one simply went half way to one's destination, climbed into a little boat, and switched to another ship.

The mind is like that: It can throw up blinders in response to stress. The Edith Russell syndrome takes many forms, ranging from Pliny the Younger sitting down to read a book while Vesuvius shook his house apart to the man who makes jokes as he goes to the gallows. Such acts may seem like a joke of nature—pathological, even—but they are neither; they are a by-product of our will to live, a physiological reaction with survival advantages.

Imposing a mental gag order upon ourselves in the midst of disaster allows us to focus, at least in part, and for a little while,

214

on something other than the frightening realities of lava or sinking ships. A certain amount of denial and detachment from extraordinary events makes them seem momentarily less extraordinary, less real, easier to handle. The gag order probably works best if attention can be diverted to something familiar and comfortable— and what is more familiar than home, or the everyday workplace? Hence, the nesting instinct is summoned into action, to create the illusion of normality and make the unbelievable tolerable. Later, when you are safe, the blinders can snap down and reality can seep in. Distanced from the danger, it does not matter if you begin to realize with fresh horror how truly great the danger was. It no longer matters if you begin to shiver, even to the point of becoming nonfunctional.

Of course, the instinct can sometimes go awry, somewhat like a miswired signal, and actually block reality out altogether, as when several passengers stayed aboard the warm and familiar *Titanic*, watching lifeboats descending half empty on the tackles, denying that the ship could possibly sink. A small minority of Icelanders would undoubtedly have continued sweeping their houses until the lava engulfed them, and a similar minority at Love Canal has refused to leave the nest. Mental gag orders, when they become denial carried to an extreme, can be just as debilitating as outright panic.

Edith Russell has provided one of history's most dramatic accounts of the gag order. We have further illustrations from Iceland and Love Canal; and most of us can probably recall instances when we have responded to major events in our lives in a seemingly inappropriate fashion. But this is by no means exclusively a twentieth-century phenomenon. There are indications, in the ashes of Thera, that the Minoans of 1650–1600 B.C. behaved similarly.

Five thousand years . . . fifty thousand years
. . . five hundred thousand years. . . . If you
free yourself from the conventional reaction to
a quantity like a million years, you free yourself
a bit from the boundaries of human time. And
in a way you do not live at all, but in another
way you live forever.

—John McPhee, *Basin and Range*

To historians: After studying the answers to all
those questions given by vulcanologists and archaeologists, do you really think that time is
the same thing for both of them?

—H. van Effenterre at the
Athens Thera Symposium, 1980

QUEEN HATSHEPSUT: THE DIFFICULTIES OF DATING OLDER WOMEN

SUMMER, A.D. 1970

Spyridon Marinatos did not sleep easily. The LM1A-LM1B discrepancy would not let him. At least twenty different pieces of a puzzle were turning round and round in his head, and it was becoming impossible to fit them together, because some of the pieces kept changing shape.

There was, for example, the problem of the squatters, as he came to call them: people who had dwelled for a time on the southern edge of the abandoned, earthquake-devastated city. There were indications that grasses and weeds had begun to grow on top of the ruins, and that dark, loamy soil was accumulating at the bases of jars standing upright on Telchines Road. Marinatos's squatters had shoveled away some of the new soil, meaning that the loam had formed before they arrived, and that at least one, possibly a dozen or more rainy seasons had passed between the earthquake and the coming of the squatters. But there was, as yet, no way of measuring the time span. All that could be said was that an earthquake had cracked stone staircases in half and overturned walls. Someone in West House had apparently attempted to replaster a wall over the second-story toilet. The two jars of plaster were still there. Someone else had set the man-sized storage jars upright near Delta House. And then, for reasons that Marinatos suspected might never be discovered, the original owners of West House, Delta House and all the other buildings thus far explored hauled up stakes and left with their valuables, leaving no indication that they planned to return.

There was no telling who the squatters were or when they came and left. They could equally plausibly have been looters, immigrants hoping to rebuild the city, or a shipload of marooned sailors. Some archaeologists have assumed that they arrived shortly after the earthquake and abandonment, and were driven out shortly before the burial of the city and the final explosion. There was, of course, the point that a quantity of food still remained in the storage jars on Telchines Road. And the squatters, whoever they were, had not accomplished very much. They'd managed to clear rubble from a few roads, to erect battering rams for the demolition of walls (they are still there today, beside the battered walls), to set up campfires outside the buildings and to convert a parlor into a workshop. From these things, Marinatos's colleagues had put two and two together and arrived at what seemed to him a perfectly reasonable but totally misleading answer: the squatters had stayed only for a couple of months, and therefore the interval between the earthquake and the burial of the entire city under a shroud of pumice must have been short.

Marinatos could not see it that way. While it was true that they had stayed only a short time, there was no reason whatsoever to assume that they'd arrived shortly after the earthquake and left shortly before the burial, or to conclude from those assumptions anything about the interval between earthquake and burial. So the interval shifted relentlessly in Marinatos's mind, from months to years, to decades and back again to months.

He was convinced that the only thing anyone could be reasonably certain about was that the work done by the squatters was within the scope of a dozen men toiling for about two months, and that at least two months had passed between the quake and the burial. Add to this the fact that from all indications the city had been large, yet the area excavated was, by comparison, pathetically small. There was no way of knowing whether traces of squatters in two buildings south of Delta House were exceptional or characteristic of what lay beyond.

And who were these other people, anyway? Could it be, Marinatos wondered, that they were treasure hunters? Could it be that the volcano had sent them scurrying before they found what they were looking for? Marinatos liked to think so. Apparently, the squatter activity had been concentrated in the southern part of the excavation, close to where the harbor must have been, precisely where you'd expect a ship's company to land and first occupy. And most of their activity seemed to have been focused around knocking down walls with stone anchors transformed into battering rams—as if they were searching for something.

It was on sleepless nights such as these, when the uncertainties turned outward and outward and made him uneasy, that Marinatos took solace from the stars. He did not use his telescope nearly so much as he had during the midsixties, before Thera came into his life. The excavation was consuming all of his time, consuming him, and for months at a time the telescope was packed away in boxes. But tonight was as good a time as any for astronomy, his second love.

More important, tonight was hauntingly appropriate. That afternoon he had been walking down from the village of Akrotiri when something shiny jutting out of a new roadcut had caught his eye. Embedded between two ash layers was a sliver of glass barely more than an inch long. Picking it up, he discovered that it was not glass at all, but a piece of clear crystal quartz that someone, long ago, had cut and polished. Marinatos tried to reconstruct a sculpture from the fragment in hand, and in his mind's eye the sliver became a cross-section through a disk, convex on both sides . . . about two inches in diameter . . . lens-shaped . . . Lens? And then, in a flash of abstract fantasy: a lens—a magnifying glass? Eye glasses? A telescope! *A Minoan telescope?*

But, no. That was impossible. At any rate, there was no way of ever proving it.

So he stowed the fantasy away in favor of present reality. The reality was that digging into the Atlantis legend was bombastic enough for one man's reputation. Most scientists liked to regard themselves as members of an elite priesthood. There seemed to be a tacit equation: the more interesting your subject was to the public at large, the less seriously your fellow scientists took you or your work. Atlantis was beyond interesting; it was the stuff of science fiction. Some likened it to looking for whatever killed the dinosaurs, or to listening for intelligent signals from outer space. It might be rip-roaring fun, but it was not respectable.

Marinatos did not have many choices. More than his image was at stake. Conducting science with the public looking on—what some called science in a fish bowl—could both help and harm the site. He could, for example, call in all the major TV networks and talk at length about the lost city of Thera, thus firing the interest of millions of people, and acquiring even more government funding for the excavation; but he would, at the same time, be inviting millions of tourists who would erode the site with their shoes and occasionally spirit away a brick or a piece of pottery. The stream of people would become a constant and unwelcome interruption to the progress of the excavation, especially during the summer months, in which the tourist season would coincide with and hinder the digging season.

The hazards of publicity were endless, and far outweighed any advantages. The potential wealth to be gained from the site might encourage the Ministry of Tourism to take charge; it was even conceivable that, in an effort to make the buried city more attractive to visitors, the ministry would decide to spruce up the buildings by reconstructing them, by removing them from their true historic context of a city wrecked by an earthquake and abandoned. Bureaucrats did not understand that they'd be ruining, beyond all possible recall, much that the ruins could tell archaeologists.

So Marinatos decided to keep a low profile. He'd continue to tell groups of school children in Akrotiri about the city beneath their island. He'd even give an occasional interview for educational television stations. And that would satisfy the showman in him. No harm would be done; everyone would be happy.

And never, never would he make public any wild ideas about ancient lens cutters. If, as was usually the case with pet theories, this one turned out to be wrong, the ridicule from peers could cost him this site, and clear the way for Doumas to succeed him. No, nothing short of a fully preserved telescope buried in a corner of West House would permit him to make public any of his musings about Minoan astronomers. Barring such a discovery, the sliver would remain just another obscure, unidentifiable and rarely talked about oddity—nothing more.

So it seemed appropriate to Marinatos, on this summer night, that he dust off his own telescope and point it in the general direction of Sagittarius. Publicly, he would remain mute about his speculations. Privately, he could dream as much as he wished.

All his life, dreams had been his friends. So the dreamer stood alone atop a buried city, just Spyridon Marinatos and the inky dark above, through which stars burned fiercely bright, yet cold, across the intervening light years. He aimed his scope at a part of the sky that included the star Ross 248. There was not much to see, nothing except distant suns, one gold, another blood bright and another sapphire pink. No one had seen their attendant worlds, but they were there, undoubtedly. Every star in the sky was probably another solar system, and the stars were so numerous that most astronomers agreed the existence of life out there was a statistical certainty. Ross 248 was the star toward which the *Voyager 2* spacecraft, now under construction, would drift after it completed a grand tour of the solar system in August 1989. During the spring of A.D. 42,165, *Voyager 2* would pass within 1.7 light years of that alien sun, carrying with it laser disk greetings, pictures and songs from earth.

Thinking about the advances the Minoans had made, Marinatos could not help wondering if actual colonists from earth might already have crossed the ocean of interstellar space, if only the civilization beneath his feet had not been living in the wrong corner of the wrong sea when the mountain exploded, the fires roared, the earth darkened and chilled, the waters were poisoned.

And when might that have been? Marinatos wondered. His own personal hunch was 1450 B.C. Some archaeologists were putting their money on 1200 B.C., and the latest carbon-dating results from the charred trees at the bottom of the Fira quarry suggested 1640 B.C.—but that was impossibly old, wasn't it?

Marinatos envied the astronomers their certainties. Their world, their understanding of time, seemed far more exact, if somewhat bizarre. An astronomer could tell you, for example, that when *Voyager 2* leaves the solar system at nineteen miles per second, it will be traveling forward through time at a rate slightly slower than the one second per second measured by the clocks on Thera, or anywhere else on earth. At the end of one of our terrestrial years, 365 days minus one half second will have passed aboard *Voyager 2*. In other words, the spacecraft, according to Albert Einstein's special relativity theory, will be moving a half second into our own future during each year of its journey, and when it reaches the star Ross 248, it will be nearly six hours younger than the planet it leaves behind.

That was the refreshing thing about astronomy and physics; they were, by comparison to archaeology, orderly and predictable. So much about Thera was unknown, and might forever be unknown. Perched as he was between the city and the stars, standing alone atop a dead world, Marinatos realized that his road would be long and hard, for it was paved with uncertainties.

For a start there was that damned LM1A-LM1B problem.

To establish a chronology of eastern Mediterranean history, archaeologists had relied mainly on an analysis of changing styles

Much as the changing styles of Coca-Cola bottles (for example, from glass to Styrofoam and glass to plastic) might one day be used to date land fills, burned and abandoned buildings, and other archaeological sites of the twentieth century, changing pottery styles are now used to date ancient ruins. A most important time probe into the last days of Thera is the transition from Late Minoan 1A pottery (*left*, characterized by lines, swirls and geometric shapes) to Late Minoan 1B (*right*, characterized by animal motifs, typically dolphins, fish and octopi, for which it is often referred to as the "marine-style" pottery).

of pottery, and on a census of other objects imported and exported to distant lands. The Cretan pottery style known as Late Minoan 1A, characterized by geometric shapes girdling a pot or vase, was found everywhere on Thera and Crete, and inside Egyptian tombs contemporary with the pharaohs Hatshepsut and Tuthmosis III. It could therefore be concluded that the Late Minoan 1A pottery style (abbreviated LM1A) flourished at the time of Hatshepsut—whenever that was.

The problem with calibrating a pottery clock was that it produced only relative dates. LM1A, for example, came before LM1B, and though most archaeologists assumed the gap between the two styles to be about fifty years, no one really knew for sure. Such arbitrary gapping—of pottery style on top of pottery style, on top of scores of other pottery styles—gave an approximate date of 1450–1500 B.C. for Queen Hatshepsut and LM1A; but again, the pottery clock was a subjective instrument, not governed by the exactitudes of orbital mechanics or special relativity.

The later, LM1B pottery style was distinct from LM1A, and apparently became very popular on Crete around the time Thera exploded. Spirals, circles and other geometric shapes had been replaced by paintings of leaping dolphins, writhing octopuses, flying fish, swallows and lilies.

One of the puzzles that kept Marinatos awake at night revolved around the total absence of LM1B storage jars or vases on Thera. Assuming a fifty-year gap between LM1A and LM1B, and assuming also that the Therans were not in the habit of importing fifty-year-old LM1A pottery from Crete, there must have been a fifty-year gap between the abandonment of Thera and the emergence of the LM1B pots and vases found scattered throughout the carbonized ruins of Zakros and other Cretan settlements.

This raised two possibilities:

1. In quick succession, Thera was abandoned, buried and blown apart; but the eruption sequence did not harm Crete. Civilization progressed at a normal pace (from the LM1A to the LM1B phase) for at least one more generation before invading peoples, or something else unrelated to the Thera explosion, caused widespread fire destruction on Crete.

2. The city beneath Akrotiri was abandoned and covered with pumice and ash; then the volcano went into a phase of quiescence before exploding some fifty years later, choking, burning and drowning portions of Crete.

The second possibility was more in agreement with Marinatos's ideas about the origin of the Atlantis legend, but it conflicted with current knowledge about explosive volcanoes. The events at Mount Pelée, for example, had progressed rapidly, over a period of weeks, from earthquakes and pumice showers to the explosion that finally incinerated St. Pierre. What Marinatos did not know, what no one knew at that time, was that Mount St. Helens had buried a forest under pumice and ash in the nineteenth century, had remained quiescent ever since, and was about to reawaken with a terrific bang.

A fifty-year gap . . . Marinatos thought about that for a while. Did it have to be fifty years? *Have* to? It was difficult to say. Most of the evidence—the whole eruption sequence—was buried on the island of Thera. Putting the clues together, Marinatos added up what he already knew.

First had come the earthquake . . . abandonment . . . a growth of weeds—one rainy season, perhaps more; and the squatters, who worked for about two months: so far, at least a year. Then a layer of pumice stones, each about the size of a rice grain, covered the ruins to a depth of two inches. Then came another rainy season, during which the pumice grains oxidized on the streets and hardened into a thin layer of cement: allow another year. Then came a two-foot layer of two-inch-wide pumice stones, then a six-foot layer of even larger stones and then a pause. Roadcuts through this last layer of pumice often revealed V-shaped stream beds: therefore, allow at least one more rainy season. On the north shore of Crete, in a beach-front villa, someone had placed lumps of Theran pumice in offering cups and sealed them in a wall. From the ashes of Thera, a pumice cult seemed to have been emerging. Someone else apparently returned to Thera at almost the same time, behaving very differently from the earlier, probably long-vanished squatters. This person (or persons) constructed a masonry wall amid the half-buried buildings. The wall rests on top of the twelve-foot pumice layer, and before it was completed, fine hot ash—hot enough to crack and partly fuse the stones— buried the structure to a depth of a hundred feet. This thickest layer of ash and pumice stones defines the end of the volcano. As increasingly great amounts of matter surged out of the earth, excavating an underground chamber at least four miles wide and five miles deep, the delicate balancing act between gravity and the volcanic pressures seeking to blow the mountain apart could not possibly last. For a time, gravity's unfailing grip force-fed the eruption, yet at the same time held the mountain together. But sooner or later, gravity was bound to win. One day the roof buckled and was pulled down, the giant broke out into open air, and most of Thera exploded away from a raw, pinkish-yellow center, ripping completely apart and gushing up to the edge of space at more than twice the speed of sound.

Okay, Marinatos said to himself. The mountain probably proceeded quickly from ash to bang. And in between the quake and the ash . . . one year . . . plus two years . . . and, for the sake of argument, allow two more years between the formation of V-shaped stream beds and someone's brilliant conclusion that the worst was over, that it would be safe to come back to the island and start building a wall. Now, all Marinatos could really say for sure, given these numbers, was that a minimum of five years had passed between the abandonment of Thera and the explosion, a minimum of five years between LM1A and LM1B (assuming, of course, that the explosion had something to do with LM1B fire destruction on Crete). It might have been as much as fifty years, or even a hundred, but not necessarily so.

And thus emerged a third possibility no one else had thought of:

3. The gap between LM1A and LM1B was small, perhaps no more than five or ten years, and quite possibly less; and the volcano did indeed bury abandoned Thera at the LM1A phase, then exploded a short time later, catching Crete at LM1B.

SPRING, A.D. 1989

This is indeed detective work beyond the dreams of Agatha Christie. Working as I have these last few years with such relatively predictable things as antimatter rocket designs, I find the uncertainties of archaeology thoroughly refreshing. I like the element of doubt, the piecing together of a story, and never really knowing if you have constructed the right story; because when you get right down to it, we are dealing with human nature, not subnuclear particles that will bounce off magnetic fields and follow predictable, measurable paths. We may guess how the Therans behaved during those last days, and the ruins may provide a clue here and there, but we have far, far more unanswered questions than actual answers.

In our search for the past, I have often felt that we are like children who have happened upon a narrow opening into a vast cavern, unexplored and blacker than the deepest mineshaft. The opening is too small. We can no more squeeze through it than we can ride a time machine back to ancient Thera and have a look around. So we pull out a flashlight and peer in. Here and there the shapes are vivid and the Therans speak clearly. In the shadows, we see a refined culture in command of the sea. We get glimmers of tall buildings, bustling ports and hanging gardens. Piecing together a handful of fresco fragments, we get the portrait of a girl. She must have been about sixteen, and she was pretty, with an unmistakable look of serenity and intelligence in her eyes. We have no idea who she was, or what eventually happened to her. She could have been an old woman by the time the earthquake came, or two hundred years dead, those intelligent eyes long turned to dust. Or perhaps the painting had been completed just before the end—whenever that was. The mind loves a good mystery. Our business is to cast light into the cavern, but the details of those last days—the political intrigues, the dealings with Egypt, the exact timing of LM1A and LM1B—may forever elude us. The shadows move, yet the opening is narrow and the dark can never be fully extinguished.

Ask two different archaeologists about the LM1A-LM1B transition and you are likely to get five different opinions. We can piece the rest of the Thera puzzle together two hundred times as many ways. I know this to be true. I've been shuffling and reshuffling the pieces, drawing and erasing new connections, for nearly a year now.

As for the LM1A-LM1B problem, I think Marinatos's third

possibility comes closer to the truth than any other. A ceramic, three-legged table from Thera is decorated with leaping dolphins, looking much like the dolphins seen in a fresco of departing ships in West House. The dolphin motif is a classic feature of Late Minoan 1B marine-style pottery. We must remember, however, that despite the similarities in style, the fresco and the tripod table are not pottery. Among many hundreds of painted vases found in the buried city, not one depicts a dolphin, or swallow, or a flying fish, or anything else that characterizes the LM1B style.

Still, there are indications that LM1B pottery was about to dawn, and it seems reasonable that its emergence on Crete could have been brought about by evacuees from Thera. A swallow decorating a Cretan pot looks as if it were painted by the same hand that rendered the birds in Delta House's Room of the Lilies. I could say the same thing about dolphins on a Cretan vase. I've seen them before: on the tripod table and a wall from West House.

The idea that Therans introduced their dolphins and swallows to Cretan pottery is logical, if you consider the sudden arrival of displaced artisans into a city (Knossos) whose resident fresco painters were already gainfully employed. It is unlikely that the Therans would have been able to displace them, to wedge them out— not unless they were willing to risk making enemies of the people who had taken them in. Since the Cretan fresco market was probably already flooded, the next easiest option for the newcomers was to find a relatively open niche. The contents of ancient rubbish heaps tell us that pottery was always chipping or breaking and having to be replaced. Unlike permanently painted walls, pottery styles, like clothes, were subject to rapid change. Thus, there was always demand for newer, more attractive styles. Painting pottery might not have been as glamorous as decorating an admiral's home, but that was OK so long as it put meat on the table. (The dolphin-decorated table should also serve as a reminder that only a small corner of the buried city has thus far been explored. More objects in this style probably remain undiscovered, perhaps even a vase or two.)

Ivy leaves and swirling tendrils are a common motif on LM1A "floral-style" goblets (*left*). Nautilus shells (*right*) decorate an LM1B "marine-style" vase from Phaistos, Crete. After the Thera explosion, both motifs were adopted by potters on mainland Greece.

If my theory about Theran refugees transmitting the LM1B style from their frescoes to Cretan pottery is true, I expect that further excavations at Thera will eventually uncover frescoes of writhing octopuses. The octopus appears frequently on Cretan pottery from the LM1B period. My guess is that octopuses, dolphins, flying fish, swallows and other popular motifs on LM1B vases made their first public appearances on Theran walls.

Knowing how quickly new styles can flash through a society, and knowing also that the LM1B style was near at hand when the lost city was evacuated, the time gap between LM1A (the Theran exodus) and LM1B (a newer "marine-style" pottery, often associated with traces of Theran ash and fire-damaged buildings on Crete), can probably be narrowed, as Marinatos said it could, from one or two generations to as few as five years.

At this late stage in Minoan civilization, the Greek mainland was being drawn into the sphere of Cretan influence. The new LM1B style was so attractive, and apparently so widely accepted, that it began to be copied on the mainland, where art objects in this style have been found in Greek tombs.

And then, in a single day and night, the Minoan world, and the world itself, changed forever.

"Now, shortly after the [Thera explosion], we may observe at Knossos signs of the beginning of a new era, of a totally new political environment," writes the Swedish archaeologist Arne Furumark. "As has become increasingly clear, not only the neighborhood of the palace but the palace itself was damaged by the LM1B disaster. Afterward, repairs and partial redecoration of the walls took place. The separate elements of the wall paintings [typically of Cretan processions] are in the Minoan tradition, but the earlier rules of composition have been dissolved, giving way to principles observed in a contemporary category of pottery, stylistically .classed as Late Minoan 2A.

"It is obvious, moreover, from the existence in the palace of nine thousand clay tablets written in the Greek Linear B script and dating from Knossos' final period [up to a destruction and abandonment tentatively dated to about 1400 B.C., some 228 years after the Thera explosion], that the . . . new masters were Greeks. The change in administrative language and script, Old Cretan and Linear A being superseded by Greek and Linear B, is indeed highly significant. The administrative system [appears to have been] a continuation of the old Minoan tradition, but its organization and contents are different, actually mirroring a society with several essentially mainland [early Greek] traits, but clad in Minoan dress."

Plato's Atlantis is a cautionary tale, written from a Greek perspective centuries after the fact, about a civilization that vanished at the height of its power. Chroniclers of disasters often portray the victims as evil and deserving of destruction. Logically, the victims had to be evil, or why would a just god have unjustly destroyed them? Even in modern times, the people of St. Pierre were said—after the fact—to have brought the vengeance of God upon themselves because some of them practiced voodoo. Ancient biblical texts hint that the Minoans (Philistines) deserved destruction because they were pagans who worshiped bulls, and allegedly had orgies in the streets. It is probably safe to guess that Sodom and Gomorrah, if they existed at all, were fairly boring cities before tales of their destruction began to spread.

So, in Plato's vision of a rational and just universe, an Egyptian memory of contact between Greece and the Minoans became the story of a once refined Atlantean culture transformed into an arrogant, evil empire, which came into violent conflict with mainland Greeks, and (with a little help from Zeus) was ultimately defeated by them. The archaeological record suggests that the Minoan world was indeed occupied by mainland Greeks near the time of the Thera explosion. Plato's version, self-glorifying as it is of Athens, does link the conflict with the time of a great natural disaster:

On the lost island of Atlantis, they possessed true and in every way great spirits, uniting gentleness with wisdom. . . . They despised everything but virtue, caring little for their present state of life, and thinking lightly of the possession of gold and other property, which seemed only to burden them. . . . But when this divine portion began to fade away in them . . . and human nature got the upper hand, they then, being unable to bear their fortune, behaved unseemly, and to him who had an eye to see, grew visibly debased, for they were losing the fairest of their precious gifts; but to those who had no eye to see true happiness, they still appeared glorious and blessed at the very time when they were becoming tainted with unrighteous avarice and power. Zeus, the god of gods, who rules according to law, and is able to see such things, perceiving that an honorable race was in a woeful plight, and wanting to inflict punishment on them, that they might be chastened and improved, collected the gods into their most holy habitation. . . . (*Critias*, 120e–121d)

Now . . . Solon . . . the island of Atlantis . . . had subjugated the parts of Libya within the Pillars of Hercules [Gibraltar] as far as Egypt, and Europe as far as Tyrrhenia [Italy]. This vast power, gathered into one, endeavored to subdue at a blow our country [Egypt] and yours

[Greece], and the whole of the region within the straits [the Mediterranean*]; and then, Solon, your country shone forth, in the excellence of her virtue and strength, among all mankind. . . . She defeated and triumphed over the invaders, and preserved from slavery those who were not yet subjugated, and generously liberated all the rest of us who dwelt within the Pillars of Hercules. But afterward there occurred violent earthquakes and floods, and in a single day and night of misfortune all your warlike men in a body sank into the earth, and the island of Atlantis in like manner disappeared in the depths of the sea. (*Timaeus*, 24d–e)

Greek conflict with an island civilization . . . the sinking of the island . . . the simultaneous destruction of a fighting force (including, evidently, Athenian soldiers) . . . tsunamis—it's all there. The story may be slightly out of sequence (the archaeological record suggests that Greek forces moved in after the disaster, not before it), and perhaps a little garbled (the distinction between Thera and Crete was apparently lost on the Egyptian priests), but all of the essentials are there. Archaeology, geology, mythology and history are, to one degree or another, converging and agreeing.

I doubt that Spyridon Marinatos appreciated this, back in the summer of 1970, but as he uncovered his lost city, he was opening up a grand arena of interdisciplinary cooperation. It was like a gigantic spider's web: pluck a thread in one corner, and it resounded somewhere else, sometimes in the most unlikely of places. Suddenly, everything about archaeology and the Minoans was becoming hitched to everything else in the world.

From the historians came ancient legends of unusual and terrifying darkness in the eastern Aegean, but apparently not in the

* For a civilization said to be coming out of the Atlantic, the pattern of conquest is strange, moving, according to Plato, from the eastern Mediterranean shores of Greece, Libya and Egypt to the rest of the Mediterranean—toward, not from, Gibraltar—with no mention of conquest in northern Spain or along the Atlantic shores of Africa. Plato has described a civilization radiating out of the eastern central Mediterranean, from the direction of Minoan Crete.

west. From oceanographers came samples of ash that had spread from Thera east across the sea, and southeast as far as Egypt, where a biblical tradition also tells of a thick darkness throughout the land, a darkness that could be felt. The death cloud deposited a dense ash layer hundreds of miles east of Thera, but penetrated west only sixty miles, stopping at the island of Melos. To halt the cloud at Melos, the headwinds from the west must have been very strong, and from the meteorologists came word that westerly squalls were almost exclusively a September through November phenomenon on the Aegean, suggesting that the volcano exploded in autumn.

From Thera itself came carbonized tree trunks, still rooted in Minoan soil at the bottom of the Fira quarry. The dead trees have provided us with a clock through archaeological time. Carbon-14 is produced by neutrons from the sun showering down through the atmosphere. These collide with abundant nitrogen nuclei in the air we breathe, chipping off a positively charged proton and replacing it with an uncharged neutron of approximately the same mass. The nitrogen is thus transformed into a heavier, unstable (that is, radioactive) variety of the carbon atom. It achieves stability by decaying back into normal nitrogen again.*

Meanwhile (this does get complicated), plants and animals are unable to differentiate between the unstable carbon-14 atom and the normal carbon-12 atom, and utilize both with equal ease. Death ends this process; an organism's carbon-12 remains, but the carbon-14 changes back into ordinary nitrogen at a predictable rate: one half will disappear every 5730 years. The Theran tree trunks could thus provide a date for the burial of Minoan settlements. If, for example, the ratio of carbon-12 to carbon-14 sug-

* A normal, nitrogen-14 atom has seven protons (+ charge), seven neutrons (no charge) and seven electrons (− charge). A normal carbon-12 atom has six each of protons, neutrons and electrons. The unstable carbon-14 atom has six protons, eight neutrons and six electrons. It achieves stability by a process called beta decay, whereby a high-speed electron is ejected from the nucleus via the transition of a neutron to a proton, which raises the atom's atomic number, or + charge, from six to seven and changes it into nitrogen-14 again.

gested that exactly half of the carbon-14 had been lost, the trees would then have died around 1760 B.C.

The carbon-14 verdict from the Fira quarry trees was that life on the island had ended in the seventeenth century B.C.—about 1640, give or take thirty years in either direction.

The conventional date, derived from the pottery clock, placed the LM1B period and the burial of Minoan Thera about 1450–1500 B.C.

Could the pottery chronology be off by almost two centuries? Marinatos didn't think so. He decided that the paleontologists and the nuclear physicists were mistaken. He wouldn't believe it. To this day, many archaeologists don't believe it. They are sticking to their guns on the pottery chronology, insisting that we paleontologist-types are full of hot air, that our way of dating things must be in error.

Personally, I don't understand what all the fuss is about. The pottery dates were from the very start arbitrary, based on an (itself arbitrary) estimate that fifty years had passed between each new pottery style. Dozens of different styles were then stacked on top of each other to derive dates. Unfortunately, the dates began to enter textbooks, and each new class of archaeology students was required to memorize them. From that moment they became self-perpetuating dogma. When some of those students became practicing archaeologists, there arose a tendency to view the dates as immutable fact. Encountering a discrepancy of one or two centuries was intolerable to them.

If Thera is the world's greatest whodunit (rivaled only by the dinosaur disappearing act), then these past four years have seen a furious accumulation of new clues. From paleobotanists in California and Ireland have come what, to me at least, are the most accurate and telling dates, all the more convincing because they fall within the carbon-14 range that Spyridon Marinatos found so distasteful. The bristlecone pines of California's White Mountains are known to live for five thousand years. You can count each year by the seasonal fluctuations of growth, which are laid down within the trees as annual growth rings—the feature that gives wood its grain. If you count back to the year A.D. 1816, you will find that the growth ring is heavily peppered with unusual, darkened cells, created by ice scarring during the summer season. This was Tambora's false winter, an event known from American, European and Asian records as the year without a summer. An identical frost scar appears in the 1627 B.C. growth band—a year whose summer snows could be attributed to none other than the eruption of Thera. Confirmation comes from an exhaustive study of overlapping generations of oak trees preserved in Irish peat bogs. Every century's pattern of dry and wet and somewhere-in-between summers is like a spectral signature written in fluctuating growth-ring sizes. Careful comparison allows the construction of an unbroken record of seasonal variation reaching all the way back to the last Ice Age. All trees dating from the seventeenth century B.C. display abnormally narrow growth rings sometime in the 1620s, suggesting that there were summer seasons, during that decade, too cold to support normal growth.

From the Greenland ice sheet, in layers of ice that accumulate annually, much like the tree rings, University of Copenhagen glaciologist Claus Hammer has extracted an ice layer containing traces of an acidic snowfall. It is more than three thousand years old, and though similar acid signals in the uppermost layers of snow and ice can be identified as fallout from the carbon exhaust cloud that now shrouds the earth, in the seventeenth century B.C., volcanoes were the only source of acid snow. Hammer acknowledges that ice-layer dates cannot be counted back as easily or precisely as California bristlecone rings. Keeping this limitation in mind, he places the acid snow at 1644 B.C., give or take twenty years. The age uncertainty (1624–1664 B.C.) is large enough to encompass the Irish peat-bog event (1620–1629 B.C.), the California bristlecone scars (1627 B.C.) and the carbon-14 date (1610–1670 B.C.).

So, we are probably looking at a final explosion in the autumn of 1628 B.C. Atmospheric dust then circled the globe, created a false winter during the summer of 1627 B.C. and eventually fell to earth as acid snow. If Marinatos's five-year gap between LM1A and LM1B is assumed to be true, then the buried city of Thera was abandoned in 1633 B.C., give about ten years or take two.

Christos Doumas has expressed reservations about taking scientific dating as gospel. He protests that we must consider the possibility that acid snow and the global chill of 1627 B.C. did not come from Thera at all, but from some other volcano.

Deep-sea research refutes this theory. Research vessels have collected bottom samples from the seven seas, and across the lengths and widths of the four oceans. Volcanic explosions of the magnitude required to create false winters distribute detectable ash layers over thousands of miles of seabed, yet no other candidate for the 1627 B.C. freeze has turned up.* The California bristlecones record only two other frost-ring events during the first three millenia B.C., and both are easily identifiable in ocean sediments. The first freeze seems to be linked with the 1900 B.C. explosion of Mount St. Helens, only a few hundred miles north of the trees. The second frost-ring event follows, by a year or less, the 44 B.C. eruption of Mount Etna, and also coincides with an acid signature in Greenland ice dating to about 50 B.C. (give or take twenty years).

One other piece of the jigsaw puzzle, only recently brought to light, provides a glimpse of life under the pall of volcanic winter, and can be dated with reasonable assurance to the time of the Thera upheaval.

In China, during the Minoan Linear A Period, records were written on strips of bamboo. Fourteen hundred years later, all such strips that had survived to the reign of Emperor Qin were

---

* For trivia enthusiasts: The largest volcanic explosion in the geologic record (judging from the thickness and extent of its deep-sea ash layer) occurred in New Zealand about 700,000 B.C. It blew an enormous hole in the earth, known now as Lake Taupo, which despite nearly three quarters of a million years of sedimentation, is still twenty miles in diameter. The next largest explosion was Toba on the Indonesian island of Sumatra in 71,000 B.C. In the mountain's place stands a volcanic crater lake twenty-four miles in diameter.

recognized as priceless treasures and, probably under orders from the emperor himself, were compiled and copied by scribes—copied so many times that survival of China's past into its future was virtually guaranteed. The ancient texts state that "in the twenty-ninth year of King Chieh [the last ruler of Hsia, the earliest recorded Chinese dynasty], the Sun was dimmed . . . King Chieh lacked virtue . . . the Sun was distressed . . . during the last years of Chieh ice formed in [summer] mornings and frosts in the sixth month [July]. Heavy rainfall toppled temples and buildings. . . . Heaven gave severe orders. The Sun and Moon were untimely. Hot and cold weather arrived in disorder. The five cereal crops withered and died."

The bamboo annals further record that floods and ice were followed by seven years of drought lasting into the beginning of the Shang dynasty. A great famine broke out, and in the northern provinces man became a maneater.* The Chinese scribes did not provide precise dates for these events; but they did footnote their royal geneaologies with listings of eclipses and other astronomical phenomena. During 1990 and 1991, NASA/JPL astronomer Kevin Pang carefully tracked China's dynasties backward through time, using as probes the predictable motions of heavenly bodies to derive such precisely dated events as the lunar eclipse of January 29, 1137 B.C.—which, though not dated by the scribes, was said by them to have occurred during the thirty-fifth year of King Wen.

King Chei lived at the same time as T'ang (the first king of the Shang dynasty), which, according to the scribes, was sixteen generations before King Wen. Because the Chinese considered a generation to be thirty years long, one can infer that Chieh ruled about 480 years before Wen—around 1617 B.C., plus or minus a

---

* The climactic disorder described by Chinese scribes was probably the result of a complex and largely unpredictable reshuffling of trade winds and ocean currents that will accompany any large scale warming or cooling of the earth. It is, perhaps, a portent of things to come if the current accumulation of greenhouse gases is allowed to continue unabated.

decade or two. Armed with additional eclipse dates for 1876 B.C. (twenty-five generations before Wen) and 1302 B.C. (five generations before Wen), Kevin Pang plotted the eclipses on a graph, fitting a curve through them and locating the point that, according to Chinese history, places Chieh sixteen generations before Wen.

"We find the date is again [in the range of] 1600 B.C.," says Pang, "plus or minus thirty years. Thus the historical records confirm what was suggested by the ice cores, tree rings and older radiocarbon dates—that Thera [exploded] late in the seventeenth century B.C."

Given such evidence, I think the archaeologists will eventually come around. One thing is certain: these are exciting times for astronomers, glaciologists and paleontologists to be poking our noses into the field. Resetting dates tends to ruffle a few feathers, but Cretan and Theran scholars haven't put up any serious resistance. People have not been studying the Minoans for very long: the civilization was only discovered during the past century, and there has not been enough time for opinions to become deeply entrenched.

Ancient Egypt is quite a different matter. Egyptologists have been studying temples and tombs for generations and generations, and so intensely specialized are they, often dedicating their lives to a single pharonic dynasty, that geologic time may lose its meaning to them. The only archaeologists I have encountered who sometimes have difficulty believing that the earth is more than sixty thousand years old have been Egyptologists. Suffice it to say that one Egyptian scholar became so disturbed by news that some of his pottery dates may have to be rewritten that he began to confide in me some chillingly detailed suicide fantasies. Since I was depending on this man to get me out of the desert alive, I decided not to press the issue. As far as I can recall, he is the only person ever to have succeeded in shutting me up.

Oceanographer Daniel Stanley has some exciting new insights on paleontology and the pottery-clock discrepancy. He and Harrison Cheng, both of the Smithsonian Institution, discovered a Nile Delta ash layer and have identified its chemical fingerprint as an exact match with ash from Thera. Equally important: the ash dates within the same ballpark as the California bristlecones, the Irish bog trees, the Greenland acid layer, China's summer frosts and Theran carbon. Of course, one cannot date the ash itself. The argon isotopes native to the ash decay so slowly that changes are discernible only across scales of tens of millions of years. We have to rely on the more rapid decay rate of carbon-14, which you don't find in the ash layer. But there is dead organic matter in the mud upon which the layer fell, and in the mud that was later deposited on top of it.

"So we have the ash layer bracketed by radiocarbon dates," explains Stanley. "The mud tells us that the ash fell around thirty-five hundred years ago. If you consider that worms and other organisms were drilling up and down through the layers, stirring up the rotting leaves and reeds we are reading from, then our age uncertainties must allow for a hundred years or more in either direction. Mud is nowhere near as accurate as a California pine, but I think we're in the same time frame as the pine-tree freeze, within the errors of C-14."

"And the Exodus connection?" I asked.

"That's of course the interesting thing. I mean, what else can you say? Hail mixed with flames? A darkness so thick that people could not see one another? A pillar of fire lighting the night sky? The Bible makes reference to all of these things, and all of these things could be descriptions of the ash cloud. This Nile ash layer is the first hard proof of anything like that in Egypt. We now have a record that statements given in Exodus which sound very strange most likely did happen. *Something* happened . . . at about the right time, I suppose, for some archaeologists. And others, who have their minds made up on one pharaoh or another being the pharaoh of the Exodus—" Stanley broke into laughter. "Well, they may not like it, but finally we've fixed a hard date on it. There it is."

Now, who was ruling Egypt around 1628 B.C.? That's a tricky one, "a sticky wicket," as Stanley puts it.

We have to begin by forgetting everything the Egyptologists "know" about dates derived from the often disputed (within circles of Egyptologists) comparison of art designs with Egyptian pharaonic dates (also hotly debated).

Next, we look for traces of Egypt in the Minoan world, starting at Thera, where some of the walls are decorated with a rare, copper-calcium-silicate pigment called Egyptian blue—which was developed and produced on the Nile and undoubtedly exported to Thera and Crete. But the blue pigment tells us only that contact with Egypt had been established at some time before the coming of the ash layer. Our first real clue as to whose dynasty we can place near this layer comes from Crete.

After the appearance of LM1B pottery and the partial rebuilding of the palace at Knossos, one administrator or another built himself a tomb on the outskirts of the capital. When he died, the tomb was filled with objects from his life, including clay tablets covered with the early Greek, Linear B script, examples of marine-style pottery and an alabaster vessel imported from Egypt and inscribed with the name of Tuthmosis III, who ruled for a time in a coregency with Queen Hatshepsut, history's first woman pharaoh.

Queen Hatshepsut was said to have been a divinely born savior.* For approximately twenty years she maintained majority control over her stepson, Tuthmosis III, apparently concentrating

---

* One of the pharaoh's functions was to be judged in heaven after death. Sooner or later (so the story went), a true pharaonic savior would be resurrected, returning to earth on judgment day to guide the souls of Egypt into a better world. When scholars began translating hieroglyphs during the nineteenth century, they discovered that ancient Egyptian voices had been speaking to us all along, through the traditions and beliefs of western religion. In addition to judgment day, saviors and resurrections, there were descriptions of baptisms, holy communions, the four horsemen of the Apocalypse and snakes in the garden of Eden, whom the heavens punished by removing their legs and making them crawl on their bellies. Apparently, when our ancestors invented new religions, they did not want to stray too far from what everyone else already knew—not if they expected to find willing converts.

on the peaceful aspects of rule rather than expanding the empire into neighboring territories. Her stepson eventually succeeded her, and one of his first motions was to have the face of her golden death mask chiseled away from every layer of her sarcophagus (which was composed of several body-shaped boxes fitting one within the other—one of which survives, in part, in the Cairo museum). He then declared the rule of a female Horus blasphemous, and ordered almost every trace of Hatshepsut's reign destroyed. Next, he expanded the empire, making war with his neighbors and leading at least one military campaign into Canaan, the Promised Land.

If a man named Moses really did exist at the time of the ash layer, then Hatshepsut was not the only royal blasphemer against Egyptian traditions. The Bible itself tells us that Moses had connections (possibly blood ties) with the pharaonic family. We can speculate endlessly about the similarity between the names Moses and Tuthmosis (was the former a derivative of the latter?); but at this writing we can do little more than hypothesize—except for one thing that I find particularly interesting. It is significant that within a century of Tuthmosis III's reign, the heretic pharaoh Amenophis IV (also called Akhenaton) made official the belief in one god, one creator. His successor, the boy-king Tutankhamen, quickly put an end to that particular line of thought. The years surrounding the ash layer were evidently ripe for heresy.

Meanwhile, back to the business of tossing out the old calendars, and calibrating time all over again, beginning at the formation of the Theran ash layer in the autumn of 1628 B.C. In the tomb of Semut, architect and vizier to Queen Hatshepsut, one wall is a fresco of men carrying Late Minoan 1A pottery vases. The men in the procession are wearing kilts identical to those worn by men at the helms of ships painted on a wall of Thera's West House. The implication is that the Semut fresco was completed during Crete's LM1A period, before Thera exploded. Therefore, at least some portion of Hatshepsut's reign must have preceded the ash layer.

Men wearing Minoan clothes are next depicted in the tomb of User Amon, a vizier to Tuthmosis III about twenty years after the death of Semut. One of the Minoans is carrying a rhyton (an ornamented scoop) shaped into a bull's head. A ceramic scoop very much like this has been recovered from the ruins of Thera. An inscription in the Egyptian tomb refers to "gifts from the islands of the Great Green." It seems likely that the vizier sailed personally to at least one of the islands, as the ruins of Knossos bear a broken statue inscribed, in Egyptian hieroglyphs, with User's name.

A third, and later, procession fresco is the most important one. It is located in the Karnak tomb of Rekhmire, one of the last viziers to Tuthmosis III. An inscription identifies the fourteen men shown in the painting as visitors from Keftiu. Most of them are carrying pottery vases painted in the LM1A style, but one man holds an LM1B rhyton, indicating that Marinatos's pottery transition has taken place.

The pottery is not all that has changed. The Rekhmire figures were originally painted wearing the short, simply patterned kilts seen in the Semut, User and West House frescoes. We know from a newer, second layer of paint (which has partly flaked away from its undercoat) that someone repainted the figures to show longer, more elaborately decorated kilts identical to those worn by men painted on pottery vessels from mainland Greece, and also on the palace walls of post-Theran Knossos.

Now large pieces of the puzzle are fitting together—fitting almost too well. The Rekhmire fresco records the transition we are looking for and allows us to place the ash layer in the reign of Tuthmosis III. The fresco tells us one more thing: whatever the consequences of the Thera explosion, it did not sever trade between Crete and Egypt. Within a single generation, between the year Rekhmire ordered the construction of his tomb and the year it was sealed with him in it, Thera exploded, the LM1B style emerged, and Cretan Minoans became subservient to mainland Greeks. Shipping might have been temporarily interrupted, but

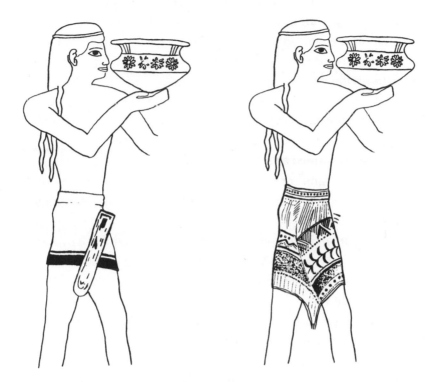

A man bearing gifts from Keftiu (Crete) was painted on the interior of
the tomb of Rekhmire, a vizier to Tuthmosis III. The man was originally
painted wearing a traditional kilt and codpiece (*left*), but was overpainted
(*right*) during Rekhmire's lifetime to show a longer, more ornate kilt
characteristic of mainland Greeks, who began to dominate the Minoan
world in the aftermath of Thera. By ordering kilts to be repainted,
Rekhmire, as Egypt's foreign minister, was giving diplomatic recognition
to a change of regime at Knossos. The painting suggests two things: (1)
The change of kilt style and regime, and hence the Thera explosion of
1628 B.C., probably occurred during the reign of Tuthmosis III, and (2),
whatever the consequences of the explosion, it did not completely snuff
out the bright flame of Minoan civilization. This is in agreement with
what the Egyptian priests of Neith told Plato's ancestor Solon (*Timaeus*,
23c): that Greek civilization could be traced back to a few learned survivors of the Atlantean deluge.

within a decade or two, people who looked Minoan, yet wore kilts that were not quite Minoan, began arriving in Egypt.

This interpretation of the Rekhmire fresco is by no means original to me, though I ally myself fully with it. Credit for the connection belongs to Trinity College historians J. V. Luce and F. Schachermeyer.

"Rekhmire, as vizier, had the duty of receiving vassal princes and their offerings," explains Luce. "He could be regarded as Egypt's foreign minister. The change in his tomb decoration can hardly be discounted as a mere artistic whim. It must surely have had political significance. By ordering kilts to be painted over, Rekhmire was giving diplomatic recognition to a change of regime at Knossos."

The Egyptian pottery clock puts the Hatshepsut and Tuthmosis III coregency between 1500 and 1450 B.C. The clock exists, but is it keeping the right time?

Apparently not.

The ash layer, dated to 1628 B.C., runs through the Tuthmosis III period and reveals the old dating methods to be in error by about one hundred fifty years.

The transition of Cretan kilt styles from Minoan to mainland Greek appears to have occurred near the end of Tuthmosis III's reign. History tells us that his predecessor, Queen Hatshepsut, focused her attention inward, on Egypt itself—which is consistent with her new role as a pharaoh of the oppression. Tuthmosis III would thus appear to be the pharaoh of the Exodus, one of the first and most famous migrations of Hebrews out of Egypt. What is most interesting about the connection between the ash layer and Tuthmosis III is that monotheism, a central theme of Hebrew belief, would soon be adopted by one of Tuthmosis III's successors. Something very important, yet poorly understood, was happening in Egypt during the aftermath of the Exodus. I suspect that the volcanic phenomena associated with Moses' departure left a bigger impression than most people think.

History also tells us that Tuthmosis III led armies between the

Bitter Lakes into Canaan. This is consistent with biblical descriptions of the pharaoh's unsuccessful pursuit of Moses.

But the parting of the Red Sea and the destruction of the pharaoh's army by its unparting, that cannot be explained by Thera, maintains Christos Doumas. "Tidal waves from the Mediterranean could never have affected the Red Sea, but many scholars have suggested that the Hebrew words *jam suf* should be translated 'Sea of Reeds' instead of 'Red Sea.' " Professor Galanopoulos of the Athens Institute of Seismology, who has associated the volcanic eruption with the legendary flood of Deucalion in Minoan times, and the landing of a boatload of lucky survivors on Mount Parnassus, suggests that this 'Sea of Reeds' can probably be identified with Sirbonis Lake, one member of the string of so-called 'Bitter Lakes' to the east of the Nile's right branch.

Doumas adds, "The route of the Exodus, as described in the Bible, favors such an interpretation, which also sounds more logical [than running headlong into the Red Sea, only a few miles south of the Bitter Lakes*]."

Egyptologists have pointed out that rates of evaporation and refill can vary considerably on the Bitter Lakes, producing great fluctuations of water level, making it possible for people to cross in some places at certain times.

If Doumas is correct about the Exodus route, then the route

---

* The Bitter Lakes are an extension of the Red Sea. The "Reed Sea" is the northernmost of the Bitter Lakes, in fact almost a part of the Mediterranean. Some scholars prefer this route for the Exodus because the Thera explosion/Exodus connection can then be used to explain the retreat of the waters, the passage of the Hebrews across the lake bottom and the subsequent destruction of their Egyptian pursuers by the incoming tsunami. The problem with this argument is that the tsunami would have arrived only minutes after the Theran ash cloud (if not before), and if the cloud is what darkened the skies, brought about famine and disease, and frightened the pharaoh into letting Moses and his people go, there was simply not enough time for pharaoh's intimidation, and then for the Hebrews to pack their belongings and travel tens of miles to the "Reed Sea" to meet the tsunami. The Thera explosion either humbled the pharaoh with its ash or destroyed his chariots with its tsunami. One cannot have it both ways, unless the Exodus account is considered, like the Atlantis legend, a composite of more than one event whose original sequence was forgotten or misunderstood by the time it was committed to writing.

itself is shrouded in odd coincidence. During the Six-Day War of A.D. 1967, an Israeli general crossed from the opposite direction. He drove his tanks over the red sands between the Bitter Lakes to penetrate and capture the Second and Third Egyptian Armies.

When Thera exploded, it changed the history of the world. If we want to understand the catastrophe, and its cascade of consequences, we cannot limit ourselves to one small branch of archaeological research. To make sense of the puzzle, we have to look at what everyone else has been seeing, in such seemingly unrelated fields as glaciology, paleobotany, oceanography, particle physics, chemistry, classical literature, atmospheric science and ancient ship-building techniques—with a little theology thrown in. And if we really pay attention, we may occasionally think what no one has thought before.

History was always my worst subject back in high school. I found it one hundred percent boring. I wish someone had thought of beginning a history course with archaeology and the search for lost civilizations. No history teacher ever told me that history could be this interesting.

All the heavy elements that make up the rocks under our feet, the carbon and iron that pulse through our veins—all of these things were created by suns that had reached the ends of their lives and erupted into supernovae. We are all dust of the stars.

—Carl Sagan, speaking at Carnegie Hall, May 1984

We have loved the stars too fondly to be fearful of the night.

—Inscription on the crypt of two astronomers

THE
INHERITORS
OF
ATLANTIS

"As I told you before, Spyridon Marinatos was fond of the junta [in Athens]," recalls Christos Doumas. "This dictatorship was American supported . . . of course Marinatos was with them and for them." Doumas was not. Their opposing views tied in, in a very strange way, to what they discovered on Thera during the summer of 1971.

"One day we found the beginning of what we call now the Xesté 3 building, which produced quite a beautiful fresco still under restoration. The restorer shouted, 'Oh, we have ladies here!' "

The ladies wore dresses, but their breasts were bare, as was common in Minoan art. Perhaps, it was the local customs office and the painting a mere decoration. Or perhaps, being near the harbor—the archaeologists joked among themselves—it might have been a "ladies house." They thought it quite funny—all of them except Spyridon Marinatos.

"At that time, the British government was discussing with the Americans—and it was a big issue in the press—about providing harbor facilities for the Sixth Fleet. So, I made a hint to Marinatos. I said, 'Well, perhaps this was the house with the ladies providing facilities to the Sixth Minoan Fleet.' "

And Marinatos, to Doumas's surprise, replied very sternly, "These were *free* people. They did not need to provide facilities for anybody."

"I was amazed that he thought providing facilities is lack of freedom."

SUMMER, A.D. 1971

Lifting the Fresco of the Ladies from the ruins in one piece was an intricate, exhausting job, requiring total concentration. At such times, Spyridon Marinatos was impossible to live with. He was agitated to the point of panic—"Why you do that? Careful! Careful! Don't drop!"—and if he was allowed to continue spreading his panic to those around him, sooner or later he was going to cause an accident.

Doumas understood that the workers needed peace and calm. But Marinatos was the director of the excavation, and an apprentice did not simply walk up to his boss and say, "Get lost."

Doumas also understood that Marinatos fancied himself a mechanic, though everything he repaired had tended to produce spare parts. Just the same, Doumas had learned to keep repair jobs filed away in his mind for crucial moments in the excavation. So, when a worker whispered in Doumas's ear that a particularly delicate operation was about to get underway, he walked over to Marinatos and mentioned casually that the lock on one of the storeroom doors was broken. He was worried that someone might get shut inside.

"Let me repair it," said Marinatos. He hurried away to fix it, and by the time he emerged from the back rooms, Doumas and the others had finished the work.

The next day, a guard informed Doumas that suddenly every key in the compound could open the door.

"So I said, 'Professor, you know that all the keys now operate this door?' And he said, 'You only and myself know it. Nobody else. So don't tell anybody.' "

But the next day Marinatos came to Doumas and said, "You better go and buy a new lock."

"Why?" said Doumas. "Isn't this one good enough now?"

"It's good, it functions. But now it's old, and we'll take it from there and put it on the darkroom."

The darkroom door, of course, did not need a lock; and the incident told Doumas something very important about his mentor, something he would remember later when, for reasons that would never be clear to him, he found himself banished from the island. The banishment would be a mistake, but Marinatos was not one to admit a mistake, and no matter how much he eventually missed his apprentice, he'd never call him back to Thera. Ever again.

Doumas had no way of knowing this at the time, but his habit of tricking Marinatos into going away to repair a lock, or a generator, or some other valuable piece of equipment during those excruciatingly stressful moments when a wrong move might send a Minoan art treasure scattering in pieces and powder, had already saved the professor's life. Throughout the past two decades, the vessels in Marinatos's brain had grown ragged and thin. By relieving him of the unbearable tension of seeing whole frescoes lifted on ropes, Doumas had kept the blood pressure down, had prevented at least one major artery from bursting open. Even so, there occurred ominous swellings and migraines. Marinatos sometimes complained of a pain over the right eye, and likened it to a pin being driven through the socket. There was indeed a wound, not much larger than a pin hole, where capillaries had weakened and begun to leak blood.

JULY 29, A.D. 1971

Professor Marinatos looked tired as he arranged three dozen fragments of pottery on his desk to reveal a spiral, LM1A design.

He hadn't really expected anything else; but he wished that just once the right shard would turn up. All he needed was a single, unambiguous, inch-wide fragment of LM1B and he could close the time gap. We have to dig faster, he thought. If it's out there, we'll find it. We must hurry.

But not tonight. Tonight was special. He set up a folding chair outside, and brought out a glass and a bottle of Santorini's Atlantis dry white. The moon was coming up full, but with no street lights burning nearby, he could still see the dust lanes of the Milky Way.

A meteor streaked down, fiercely bright, trailing long running sparks in the direction of Crete. It stalled up there, the sparks stopped streaming off, and a tiny red dot dropped toward the horizon, dropped for perhaps a tenth of a second before cooling and fading to black.

Impact!

Marinatos thought, more than a hundred miles away, no doubt. On water. You'd never find it.

That's why he didn't dig in Athens. Too many lights. Too much smog blocking out the sky. He wondered what the Therans must have thought the stars were all about, watching from this very spot, three dozen centuries ago.

He lifted a glass toward the moon. A toast to Dave Scott and James Irwin, who are right now walking up there on Hadley Ridge, picking up rocks that have lain undisturbed for four billion years.

The moon is just a stepping stone to Mars and Europa, he reflected. There are children in this world who are going to see things of which we people barely dream. They will design new ships and sail out across the light years, into a sea of suns. So long as we do not blow ourselves up, civilization will never stand still.

That's what Thera was teaching Marinatos. How can you spend a lifetime interpreting the past without wondering about the future? We live now in what used to be the Minoans' future.

Yet under the very ground upon which they built their cities, dinosaurs once rasped and hissed. They, and not the ratlike, insect-eating mammals underfoot, had been rulers of this earth. But the sun grew cold, the climate changed, the dinosaurs vanished, and the mammals rose up to replace them. The Bronze Age inhabitants of the island built their civilization over saurian graves, then they too vanished. But it was only a little pause. What's three thousand years of stagnation and dark ages against dinosaur time?

As he sat atop a sheet of pumice, looking up at Hadley Ridge, it occurred to Marinatos that Scott and Irwin were the next step in human evolution.

"We eventually got around to where the Minoans were headed just the same," he said to himself. "A couple of thousand years earlier or later hardly matters in the overall scheme of things. Out there across the light years, time and the stars are waiting— waiting for Argosseys yet to be."

Marinatos was very keen on astronomy, recalls Doumas. And if you showed an interest in the things that fascinated him, he was like a child. He was quite willing to sit down and talk about them for hours. "He was a very nice person in company, very sociable, but very difficult to work with.

"I don't complain. I had a good time here. My bad time with him started after I left [in March 1973], because I did not want to leave. And as the head of the department [of Antiquities], he was responsible for sending me to Rhodes, where I had no scientific interest at all. Therefore, I said that he shouldn't expect anything from me anymore. And I refused to cooperate. After that, he pursued me administratively."

And what was Doumas doing there in the first place? He had plenty of time to think about that; but he could not figure it out. The banishment to Rhodes came without warning. He never told Doumas what he had done wrong. What Marinatos really thought of him, he never discovered.

By 1974 living quarters on the island had improved. Hot showers had been added to the archaeological villa, and there was a navigable road to Fira. With running water, the camp was finally up to the Minoans' living standards.

Marinatos was pushing the dig faster and faster every summer, as if he were hurrying after something. The material was accumulating at an unthinkable rate, and Doumas had been the one who'd known where everything was stored, in precisely which drawer one could find a certain pottery shard from the Room of the Lilies. The professor was becoming increasingly agitated. He was forever complaining that he could not find things, and that Doumas was the only person who knew how it had all been arranged.

Still, he was a stubborn and proud man. He never did summon Doumas back. Then the war with Turkey came and the site fell apart for a while. Athens ordered the island evacuated, but Marinatos refused to leave. Like the captain of a foundering ship, he stayed with Thera till the very end.

The island of Cyprus is located four hundred miles east of Thera, forty miles south of Turkey, and one hundred miles west of Lebanon. After the Theran tsunami leveled coastal towns there and lifted a pumice raft more than ninety feet above sea level, Egyptians invaded the island and claimed it as a naval outpost. When finally Egyptian civilization went into eclipse, Alexander the Great took over. Monarchs, crusaders and empires have been leaving their signatures upon the landscape ever since: Greek and Roman columns . . . mosques . . . churches . . . Venetian fortresses. . . . By the summer of 1974, Greek and Turkish Cypriots were living in uneasy peace. Officially, the island was free and independent of both Turkey and Greece, but every building flew either a Turkish or a Greek flag, and the Cypriot flag was rarely seen.

Against this backdrop, George Papadopoulos's military junta in Athens sent a column of tanks to the front door of Cyprus's presidential palace. They loaded, aimed and blew the place apart. Minutes later, the National Guard interrupted all regular programming to announce that the president, Archbishop Makarios, was dead and the Greek military was now in control.

The Turkish Cypriots were not about to take such news quietly, especially after President Makarios's unmistakable voice came crackling over the airwaves: "I'm not dead yet. You missed, bastards!"

Greek jets zeroed in on the source of the broadcast. They poured down napalm—and missed again. The heat-seeking missiles that leapt up after them proved to be considerably more accurate.

When the Greek fleet approached the island, it was met by Turkish dive bombers. As the first destroyer went under, Turkish paratroopers mounted a full-scale retaliatory response against Greek-held Cyprus. Then, as echoes of war resounded across the Mediterranean, Athens radio went off the air, Syria placed its forces on maximum alert, the U.S. Sixth Fleet dispatched combat vessels toward the Aegean, and the Kremlin, evidently pleased to see Turkey and Greece facing off for the first war ever between two NATO allies, alerted airborne divisions in a show of support for Turkey.

Nanno Marinatos remembers: "We were all at the excavation, relaxing one evening, when suddenly the power went out, and stayed out. Someone had a pocket radio, and from that we received the news that the island was under a blackout order. It was unbelievable news. We were told that there was great danger of war with Turkey because Cyprus had been invaded by the Turks, and Thera, being on the eastern edge of Greek territory, was a likely target. So all the islands had to be dark at night for fear of bombing.

"And then the evacuation started. A few ships would be coming to take women, children and soldiers away from the island. My father decided to continue the excavation, to continue as before, except that all of his foreign guests and main collaborators had been ordered to leave—especially the foreigners.

"So he stayed on, and I stayed on with him, the whole summer. And we continued with just a few workmen, as though nothing had happened."

Turkish jets crisscrossed the sky, keeping watch over Greek shipping. The Greeks held back, fearing that any further move against Cyprus would invite full-scale war. They had underestimated the Turks. No one in charge had believed they would fight back. By August 5, nearly two weeks into the conflict, Turkish forces had claimed one third of Cyprus as their own. Greece was losing the war. In Athens, a cluster of military officers decided that the dictator Papadopoulos and his generals had blundered so badly they simply had to go.

"Then came another message," says Nanno Marinatos. "Papadopoulos's government had fallen. That news was hard on my father, because the political implications were very bad. He had supported the old government, and when old governments fell, new governments tended to carry out purges against supporters of the old. Very quickly, we received word that the excavation would be closed down. My father was grievously offended. He took it very, very badly.

"So there were two things happening. On one hand there was fear of invasion from Turkey, and on the other hand there was political turmoil inside Greece.

"And though Athens had called him away, still my father insisted on staying with the excavation. I wanted to stay with him, but in the end he made up my mind for me. He put me on a boat and there was simply no way he was going to take no for an answer. He was like that. When Spyridon Marinatos decided something, that was the end of the argument. 'You've got a master's degree to finish,' he said. 'You've got a life ahead of you.

You have grown up with this city, I have grown old with it. Now there is only your life. Go with my love.'

"He sent me away with one of the ships. I went back to Athens, and eventually left for the United States.

"I was twenty-four years old. I did not know that I had just seen my father for the very last time."

OCTOBER 1, A.D. 1974

Now the buried city did look like a ghost town. Nothing moved except Spyridon Marinatos and three local workers from Akrotiri. A new democratic government had taken over in Athens, and the dust of the war was finally settling. Marinatos had few doubts that the new regime would soon get around to physically removing him from Thera.

He stood on a pumice ledge overlooking Telchines Road and Delta House, supervising the positioning of heavy conveyor belts, with which he hoped to expedite the removal of ash from the ruins. It was 1:30 P.M., and his lunch was not sitting well in his stomach. This only added to his agitation.

"Watch that wall. Watch out for that wall!" he shouted, and then let loose with a string of curses. An instant later, he felt something give, and pain ran through his right eye like a live wire.

The archaeologist understood all too well what was happening— a stroke. A major artery had finally let go, blown open inside his head. As the entire left side of his body went numb, he let out a cry of frustration and bewilderment. Then his left leg crumpled beneath him and he tumbled in a heap onto Telchines Road.

In seconds, the workers were ringed around him. One of them slid a hand under his head and tried to prop him up. Marinatos's right eye searched aimlessly, pausing at no one and nothing, as if he were seeing right through his surroundings.

"City," he murmured. "City . . ."

WINTER, A.D. 1975

The new government summoned Christos Doumas from Rhodes and promptly reassigned him to Thera, where he was to take Spyridon Marinatos's place.

He began by slowing down the rate of exploration. With Marinatos in charge, pumice and ash were being removed nonstop for fully three months out of every year. A month's exploration produced enough artifacts and frescoes to keep the restorers busy for more than a year. Doumas realized that he'd be working with his colleagues for the next ten years just to preserve, record and begin to interpret the material his mentor had left behind. He decided that it would be safer to proceed slowly, to record and analyze everything as it was pulled out of the ground, rather than let it stockpile in indigestible quantities.

It was ironic that when Doumas had first arrived at the site, fresh out of graduate school, his teacher had seemed the most conservative excavator in the world. In the tunnels Marinatos had carved out, there stood a doorway into a Minoan home, blocked completely with pumice and ash. Doumas had been eager to learn what was on the other side; but Marinatos explained that it would be years before anyone found out, for they must all learn to slow down. The apprentice would never have believed that a day would come when even the professor's glacial pace would seem too fast.

If there had been any uncertainty about Nanno Marinatos's direction, her father's death removed it. She decided, during the winter of 1975, that she would continue on her path toward a Ph.D. in classics, and become an archaeologist. From there, a return to Thera was only a matter of time.

Seen from a distance, at the University of Colorado, the city Nanno had grown up with was becoming increasingly fascinating to her. She now saw a genuine need to continue with her father's work, and by 1988 she had emerged as one of the world's leading authorities on the frescoes of the vanished people.

"When my father found a fleet of ships painted on the walls of West House," Nanno recalls, "he interpreted them as a glimpse into some historic voyage. He believed the ships to be heading away to Egypt or Libya, so he began looking around the site, convinced that evidence of foreign import would prove that the fresco actually depicted a trip abroad. And the search for Egyptian products in Theran and Cretan ruins did indeed prove fruitful. Minoan exports even turned up in Egyptian tombs. But as I studied the frescoes, I began to see them differently. My attention was drawn in particular to such things as streamers of flowers shown draped from the rigging. To me, it was not a voyage at all, but a religious celebration on the Aegean, perhaps a tribute to a Minoan diety of the sea.

"That is the intriguing part of my relationship with my father. He influenced me so much, and yet, as it turns out, I have often come to opposite conclusions from him."

AUTUMN, A.D. 1988

In one respect, archaeologists and paleontologists are alike: most of our work is done in libraries, laboratories and the back rooms of museums. A few have been known to get annoyed when someone at a cocktail party asks, "Been to any interesting digs lately?" I've known two extremists who actually became experts on sites they'd never visited. Back in New Zealand, where I used to be a hard-core, full-time paleontologist, I was considered one of those weirdos who spent a couple of months out of every year in the wilderness, and insisted on doing my own cleaning and repair work when I got home to the laboratory. (For some reason, this behavior is common among—almost exclusive to—dinosaur enthusiasts, who are considered a bit odd to begin with because of our subject matter and time frames.)

Thera is different. The island has spoiled me. The full extent of my own digging has been a few handfuls of ash taken from a roadcut through the pumice, just outside the roofed area. The roadcuts are the island's newest time portals through the ash layer—giant slices through Thera's skin. I took my samples between two layers of pumice laid down over the buried city, yet predating the big explosion. Nearby, the British geologist Susan Limbrex was drawing samples for later chemical analysis in her Birmingham lab. She wanted to know whether or not soil had accumulated between the pumice layers. I dumped my sample into a cup of water and stirred with a finger. Grass seeds floated to the surface. Grass seeds, unlike the grasses themselves, unlike wood, last almost forever. Even in the tomb of Tutankhamen, almost as old as Thera itself, seeds recovered by Howard Carter miraculously sprouted after thirty-five hundred years, yet everything else within the tomb showed signs of deterioration. It did not occur to me to plant the seeds from the roadcut and see if they, too, were still fertile. Their presence was enough; it meant that at least one rainy season had passed between the two layers of pumice.

"Yes," agreed Dr. Limbrex. "But was it one rainy season, or ten, or even thirty? How big is the gap between LM1A and LM1B?"

Every year she will return to the island: a day or two in the fresh air and the sun, and then a year-long retreat to poorly lighted back rooms, where the necessary hard work begins. In five years, perhaps, she will have a clearer resolution of the time gap.

As archaeological behavior goes, Nanno Marinatos is rather typical. She does not personally excavate Thera. Her work is more a synthesis of the material brought up from buried rooms than a firsthand, inch-by-inch descent into the rooms themselves.

Christos Doumas is not so typical. He, like Spyridon Marinatos before him, spends much of his time actually living at the site.

The city obsesses him, as it did his mentor. Speaking with him, you cannot doubt that he would lay down his life before he let harm come to the place. There is something distinctly territorial about his behavior, something sad. I am sometimes reminded of the butterfly fish that lay dying in the Mexican sun because it would not abandon its outcrop of coral when the tide withdrew.

SEPTEMBER 27, A.D. 1988

It was cold.

The wind kept coming in off the Aegean Sea. It skated over the corrugated tin roof and screamed thinly as it tugged at the telephone wires. The air was calm beneath the canopy, but the autumn squall was pulling up sheets of volcanic dust outside. It hung thickly over the ruined buildings and it moved around me, my wife Gloria and Christos Doumas as we returned to Delta House, and the dust got into our mouths and made our throats dry.

Ahead, near the end of Telchines Road, children were climbing on a pumice ledge.

Gloria asked who they were.

Children from the town, Doumas explained. Every day they bring lunch to their parents and brothers working at the site. "The villagers consider the site their own business, and it's very good because they take care about it," he continued. Many hands are needed to help with the mass of material removed from the site. Some of the villagers have become able excavators, digging carefully with brushes and knives. "They are very skilled and know how to behave, how to treat the finds before the restorer takes over."

But not this year. With the excavation budget cut to the bone, there is only work for guards.

We watched the children. Doumas's own children had been born while he was working at Thera. His oldest was fourteen. She was born in the year of the Turkish-Cypriot war. The second daughter was thirteen, and his son was twelve. They'd known the site from infancy. During the summers, when there had been funding and they'd joined him at the excavation, Doumas had persuaded his children to wash pottery and to sieve refuse soil for missed fragments of ancient Thera. They didn't need much coaxing. They enjoyed working with the graduate students: it made them feel that they were adults.

Being the child of an archaeologist certainly had its moments. There was the afternoon in 1986 when Doumas's second daughter came home from school and said, "Look what we have in the book of our history!" *Our* history. He liked how she'd said that. In her school book was a wall painting from West House, and his daughter explained, "I said to all the children in the class, 'We have found this; in *our* excavation.' "

"Now and then," said Christos Doumas, "we invite the school children down from Akrotiri and other villages, and I remember once, when someone asked them what impressed them most, one girl said, 'I'm impressed by the millstones because they are very similar to those we use even today,' [to make flour]."

"Yes, it's something they see every day," said Gloria.

"And another girl, who was very clever, said, 'We are proud that such a civilization came from our area; but we are very sorry that all these paintings are exhibited in Athens and not in our village.' "

Gloria and I had been discussing how wonderful it would be to have a reconstruction of West House, Delta House and Telchines Road away from the site, where visitors could see reproductions of all the frescoes. I put the idea to Doumas.

"Yes," he said, "but we would require the expropriation of another huge area like this." And that would require the expropriation of money that didn't begin to exist at the moment.

We listened to the wind. Through an open window I could see that someone had placed fresh red flowers in room Delta 16, where Spyridon Marinatos was buried. I couldn't help but wonder: with more than three hundred years of excavation ahead, how many dynasties of archaeologists would eventually leave tombs in this city?

"No," said Doumas. "An archaeologist always respects the site more than anything else. And I thnk it is a spoil to incorporate modern activity besides what is necessary for pure research. You cannot bury somebody there, especially an archaeologist."

"So Marinatos would not have wanted to be buried there?"

"I don't think so. He was very, very fond of his site, and very strict in his attitude toward the archaeological heritage. It was his wife's decision to bury him there [in Delta 16]. She said maybe it was his wish to be buried there, but I suspect that if he ever said anything, he said he wished to be buried near the site, not inside an ancient room."

Delta 16 . . . Delta House . . . West House . . . Telchines Road. . . . There are the rooms and houses, and the contents of houses, and streets. Doumas says that Telchines Road and West House just don't sound right to him. These are new names, given by archaeologists. The real names, the ones the Minoans used, are gone, lost somewhere in time. The city rose and died under those names. The new names assigned to the city and the land have little more substance than a mythical continent upon the Atlantic. They have fallen like the last spring snow upon a field.

It was easy to see how the city had died. But what about the rest of the Minoan world? Could the blast from this one island have wiped out an entire civilization? I knew that many scholars tended to think Thera was not the cause of the Minoan decline. Doumas was inclined to agree with them. He did not believe the volcano did it. Not all of it.

No, the volcano does not explain everything. But it could have been a *coup de grace*, one piece of bad luck too many, something that finally stressed the Minoan economy to the breaking point. It is easy to envision, especially in light of today's shaky world economy. Our resources are hemorrhaging into runaway weapons programs, and there is rampant mismanagement of everything else. Children on the banks of the Nile knew a far better standard of living during ancient Egyptian and Minoan times than they know there today. Electronic civilization is in enough trouble without ozone holes, acid rain, algae blooms and greenhouse effects hanging over our heads. Any of these things could become our volcano.

Doumas believes the Theran upheaval could have amplified a process, a decline that had already started for other reasons.

I thought of the proud Romans who fell, not because of invading barbarians, but because of the empire's own internal weaknesses. The barbarians simply swept in and grabbed whatever was left and could easily be taken.

Before I came to the island, I'd assumed that the dream of any archaeologist would be to see the excavation expanded. I'd assumed wrong.

"No, I don't think my life will be long enough to expand beyond the roof," Doumas told me. For the time being, he's confined his ambition to finding ways of protecting the site from earthquakes and organizing it as a museum. He's even stopped pouring wire-reinforced concrete into the hollow spaces where wooden beams had dissolved out of existence. "We'd need something light and soft—softer than the concrete, so that if an earthquake comes it can move in the same way with the rest of the wall . . . and at the same time, a material that will be able to be reinforced with steel." The archaeologists must prepare for the inevitable. With a couple of hundred years work ahead, the volcano cannot be expected to remain quiescent to the end. In an earthquake, wood will bend, and perhaps even break, but concrete with steel rods in it could kick against a wall and demolish it. Maybe plastics were the answer.

Squatting atop a broken wall, scuffing it with my feet as I tried to make a photomosaic of the "bakeshop," I began to appreciate the irony: before these walls were exposed, the surrounding ash had been a protective cocoon. They withstood thirty-six hundred years of earthquakes. Now, far more damage could be done in just thirty-six seconds. The weather, too, could destroy this city. The Minoan cements were almost as dry and crumbly as the pumice that surrounded them. A single rain could turn the walls to mud. And each year the rain threatened to get in.

Doumas called my attention to a rash of jagged holes in the tin roof.

I looked where Doumas was pointing. There was no great danger yet, but all the metallic parts of the roof were subject to corrosion, which was facilitated by the constant spray of salt air and by the sulfuric content of the volcanic dust—which was everywhere. Doumas explained that within the next five to seven years, the roof would have to be replaced. It had fulfilled its purpose—it had protected the ruins as they were first brought to light—but the difficult, expensive operation of replacing it would give him the opportunity to erect a more permanent structure with more resilient materials, and to organize the site into a museum, rather than just an archaeological dig with a rain cover.

At the La Brea tar pits, I'd told him, tourists could look in and actually see paleontologists restoring the fossils. Visitors found the paleontologists as fascinating as the saber-toothed tiger exhibit.

Doumas agreed. Tourists could view the site mostly from above, on bridges and walkways that would allow them to see into the interiors of the houses—how the bakeshop was organized, bedroom furnishings and household utensils still sitting in place, things as they were used by the Therans. And at the same time the visitors could watch the archaeologists working below. Oh, yes, he did have dreams for the site.

Marinatos's idea had been to display the city in its true historic context, as a world entombed within the earth. He'd tried to hollow out the earth's crust and display the buildings as they really were—underground—hadn't he?

But, Doumas reminded me, especially in a volcanic zone such as this, you cannot have hundreds of tons of crumbly rock hanging overhead.

I had told him of my own dreams for Thera. I saw a way both to protect the city from the elements and to recreate the atmosphere of a lost, subterranean world. I'd cover the site, not with a roof but with a two-story structure. The building would have an earthquake-resistant frame consisting mostly of tall, sturdy columns supporting the second floor. On the ground floor, where the ruins are, there would be caverns more than sixty feet high. The supporting columns, and all the rest of the framework, could be hidden behind lightweight facades, made to resemble the pumice-stone walls of the Fira quarry. Seen from within the caverns, the supporting pillars would resemble crude stone columns etched out and left behind by the quarrying process—stalagmites running from floor to ceiling. The impression I wanted to give was of a whole city quarried underground, just as Marinatos had envisioned it.

The second floor, nearly eighty feet above the city, would be open to the air, like a roof. There I'd reconstruct the buildings of the city as they once were. Using lightweight materials, mostly fiberglass and plastics modeled to resemble wood and masonry, I would rebuild West House directly over the site of the real West House, tracing its floor plan exactly but replicating its frescoes and furnishings, just as when it was inhabited. I'd do the same thing with Telchines Road, and Delta House, and all the other buildings. As the excavation expanded, I'd expand the overlying, reconstructed city along with it.

Doumas liked to think about concepts for future displays of the city, but he had to be realistic. What I proposed was expensive, and he explained that he could not be sure that they'd even give him money to fix holes in the old roof.

Here was one of the wonders of the world—or at least it was going to become that if the whole city could ever be brought to light. But for the moment, few people had even heard of the place.

When Doumas talked about funding, his eyes began to show anger. "I was not given any money this year. And unfortunately,

I learned about it in mid-May. So I had to cancel all my projects."

Looking around, I had to admit that the place was beginning to look like somewhat of a dump. The hastily constructed wooden walkways for visitors were falling apart. Empty Kodak boxes and Coca-Cola cans were accumulating. There was no one to clean them up.

"The scandal," Doumas continued, "is that they have a lot of income from the site." If the government gave him even a fraction of the money charged for admission to the site, he explained, he would be "more than well off" with the excavation.

I'd seen at least two hundred people pass through the site that very morning, and this was the off season. Doumas explained that during the summer months the site received at least a thousand tourists per day. At five hundred drachmas (U.S. $3.20) per head, that added up to at least a half million drachmas (U.S. $3,200) per day.

Counting all the workers Doumas employed, including the excavation staff, archaeologists, restorers and so on, he needed about four to five million drachmas (U.S. $25,600 to $32,000) for a month's work. That was all he had needed for the 1988 digging season: just eight to ten days' intake from ticket sales. If *all* of the intake from ticket sales were channeled back into the site, as ought to be the case, Doumas could probably build my cave and open-air reconstruction of the buried city.

"I don't know what your views are," said Doumas, "but I'm involved in the peace movement, and I know that humanity could live thousands of times better on only a fraction of the money spent on so-called armament. Armament against whom? Against humanity." He was right, of course. Our existing weapons could make the explosion here look miniature. They are enough to destroy the earth several times, if we don't pollute ourselves to death first. Just think: a new Atlantis legend . . . if anyone lives to write about it.

Atlantis . . . Without proper care, this site may actually die.

Perhaps we would have been wiser to leave the city covered up with rocks. Perhaps civilization has not matured sufficiently to receive such treasures.

"I will start from now my struggle to guarantee some money," said Doumas, and then he revealed how desperate that struggle might become. "I am prepared to expose the government publicly and internationally if they don't do anything, because it is their responsibility. They cannot only have an income from it [the site] without any attendants and repair. . . . I had a visit [from] the European parliament, and I spoke openly to them. . . . I was thinking about having an international press conference."

It occurred to me that, in telling me this, Doumas might be using me to pack a little punch behind his threats of exposure. But it did not matter. The city was all that mattered, the city above all else. Already, wood impressions had deteriorated before removable casts could be made of them. Comparative studies of tree ring growth patterns would have revealed much about the construction sequence of the buildings, but such information was lost forever. Even if funding were to be miraculously restored, the same conditions could quietly recur a year or two further on, or after Doumas was no longer there to defend the site. The public would have to know about the city's plight in order to keep an eye on it, and to make sure that this never happened again.

I began to understand why people were willing to risk their lives during two world wars to save precious works of art. As a paleontologist, I was not accustomed to coming across works of art in the ground. Yet at Thera every building was an art treasure.

An old archaeologist had once told me that it was only our art that made all the struggles of civilization worthwhile. "All our science, our technology, our mathematics—nothing is unique about them," he said. "These things will be repeated by any sufficiently advanced civilization, anywhere in the galaxy. They'll all discover that $E = mc^2$. But there is only one golden death mask of Tutankhamen, only one Room of Lilies. It is through our art that we really live and breathe. If I could pick only one thing that could survive on this earth and speak for our species, it would be our art. Now, that's all you have to remember. Only our art should live forever."

ICARUS

We detected no evidence of life. . . . We hovered over a half-buried car and tried to blow away ash with the rotors. But it boiled up menacingly. Fully two days had passed since the mountain exploded, but the ash was still very hot. I felt a growing appreciation for all of us living on a planetary crust so frighteningly afloat atop such terrible heats and pressures. Never again, it came to me then and remains with me to this day, would I regain my former complacency about this world we live on.

—photographer Rowe Findley at Mount St. Helens, May 20, 1980

Will not we fear, though the earth be removed, and though the mountains be carried into the midst of the sea; though the waters thereof roar and be troubled, though the mountains shake with the swelling thereof. Selah . . . the earth melted . . . Selah.

—the 46th Psalm

PHAËTHON
RISING

SPRING, A.D. 1989

Eight hundred miles east of Cairo, near the ruins of Babylon in Iraq, they have found clay tablets dating back to 2000 B.C. A dozen different hands etched the same assignment into a dozen slabs of clay, which were subsequently baked, meaning that they would last almost forever. Surfacing into the twentieth century A.D., the slabs teach us that six hundred years after Cheops was built, three hundred years before Tuthmosis III ruled, seventeen hundred years before Euclid introduced his "new" geometry at Alexandria, Babylonian schoolchildren were learning about the hypotenuse of a right triangle. In those same layers of earth, archaeologists have discovered little ceramic jars. They are peculiar in that they have strips of copper running up and down their inner walls, and iron rods at their centers. Several years ago, someone got the idea of filling one of the jars with lemon juice (a source of hydrogen ions, or electrons), hooking it up to wires and a meter, and seeing if a current flowed. There was no doubt about it: the thing *could* be used as a battery. But had it? For all anyone knew, the copper and iron could merely have been decorative.

Last year, during a visit to the Cairo museum, I asked one of the curators about the "batteries" and was told that several gold charms found in the tombs of lesser nobles, and once thought to have been solid gold, now appear to have been electroplated—over thirty-four hundred years ago.

Chariots of the gods?

I very much doubt it. Personally, I've always considered that hot speculation to smack of racism. No one seems to question that the Greeks and Romans were capable of inventing Euclidian geometry, steam engines, differential gear shifts and screw presses (though it now appears that such knowledge was inherited largely from Minoans, central Asians and north Africans); but mention the existence of batteries, pyramids and the smelting of aluminum in early Bronze-Age Africa—or even plumbing on Thera—and suddenly it becomes necessary to invent ancient astronauts who came down from the stars and "showed them how."

Why is it so difficult to believe that someone back in Minoan times knew a thing or two about electricity? There's nothing particularly earthshaking about it. It just means that very little is really new, and that "primitive" cultures knew more than most people think.

Which leads me to wonder how much we really do know about ancient Minoans and the last days of Thera.

Many questions remain about Minoan naval technology, and about the impact of the death cloud and tsunamis. Some think that they had hardly any effect at all, while others claim the effects were so severe that fleets and cities disappeared, and it is difficult to imagine anyone or anything surviving.

"I have trouble with the tsunamis," says D. B. Vitaliano of the U.S. Geological Survey. "It is often said that the entire Minoan fleet was destroyed. Assuming that some very destructive waves reached the ports of northern Crete, they would destroy a considerable amount of the shipping that was in the ports. But the ships at sea? The safest place to be in a tsunami is right in the open sea." Scenes like that classic one in the movie *Poseidon Adventure*, with the wave cresting over an ocean liner in the middle of the Mediterranean, simply do not happen, because, as Vitaliano explains, "it is not a breaking wave that goes sweeping across the ocean's surface; it is a gently rising and falling long wave. And when you are in a ship in the open sea, you do not even feel that gentle motion."

But if any of those ships at sea happened to be downwind of Thera, they would have capsized under the sheer weight of even two or three inches of ash on the decks. Add to this scorching and suffocation.

"I would like to comment on the possible destruction of Minoan sea power by tsunamis," said the scholar S. R. Taylor at a 1979 symposium on Thera. "We shall never know whether or not the Minoan fleet was in the harbor when tsunamis hit the coast, but a principal effect would have been the destruction of the fleet bases—including the shipbuilding and repairing facilities with their stores of masts, oars, sails, ropes, etc.—as well as any administrative arrangements. In order to repel a seaborne invasion [from, say, mainland Greece], an efficient scouting and intelligence organization would be needed to interpose the fleet between potential aggressors and the coast of Crete. Thus, the weakening effects of [Theran] tsunamis on the Minoan naval organization, rather than the destruction of individual ships, may have been a decisive factor."

Professor Rakham, of Corpus Christi College, Cambridge, adds: "The destruction of ships in harbor, by itself, could hardly have been more than a severe setback, since ships are not long-lived and would be replaced all the time. . . . [But] certainly there would have been shipwrights and ship repairers who in the nature of their business would have lived within reach of a tsunami. The drowning of these men, and the loss of their special skills, might have been a disaster more permanent than any loss of ships or equipment."

"I cannot accept any of these doom and gloom scenarios," says University of North Carolina archaeologist Jeffrey Soles. "Even a cursory examination of my excavation at Mochlos shows that the catastrophe could not have caused the downfall of Minoan civilization. For one thing, people were able to rebuild on top of the ashes; and the artists who supplied pottery for the new homes, and the architects who designed them, were clearly civilized Minoans."

Mochlos, a small island off northeast Crete, was sheltered from Theran tsunamis by a favorable incline of the shore and by a peninsula jutting between it and the volcano. During the summer of 1989, Soles and his colleagues discovered a layer of Theran ash four inches deep. Immediately beneath the layer (and within it) was the wreckage of a Minoan house, including crushed pottery in the LM1A style. On top of the ash was the floor of a new home whose U-shaped stairway was built in the classic Minoan style and whose main entrance was decorated with a slab of green limestone (a Minoan threshold stone). The next few inches of sediment recorded the eventual abandonment of the new home. The roof rotted and fell in. Hundreds of pounds of soil and generations of weeds accumulated on the rebuilt floor. Then a new family arrived, built yet another home upon the sturdy foundations of the old, laid down another new floor, and sealed beneath it all traces of the people who had first dwelt on top of the ash layer. Between this third floor and the ash, Soles found pottery decorated with paintings of nautilus shells and writhing octopi— LM1B marine-style pottery. Immediately above the ash was LM1B. Below it was LM1A. While this finding did little to narrow the LM1A-LM1B time gap (all that could be said with certainty was that, on the island of Mochlos, LM1B did not come into vogue until after the Thera explosion), what really mattered to Jeffrey Soles was that both styles were undeniably Late Minoan.

"They lived," he says. "It's quite obvious to me that Thera did not destroy the Minoans. There they are: rebuilding on Mochlos right after the explosion. As an archaeologist, I just look at the pottery styles and the buildings that follow. They are among the finest ever produced in Crete. Judging from that, I have trouble with a long-term Theran impact."

Egypt's Rekhmire fresco (which records a transition in Cretan styles from LM1A and Minoan kilts to LM1B and mainland Greek kilts) tells us that if trade between Crete and Egypt was interrupted at all by Thera, it had started up again quickly, during the term of a single vizier to Tuthmosis III. Soles is correct in his interpretation that talented artisans continued to find work after the explosion, and it is clear that trade routes remained active; but what does that really tell us?

One can stroll through midtown Manhattan today and see evidence that architects of the 1930s were able to find work and in fact directed the construction of the most beautiful buildings ever produced in the city. Without written records of those times, an archaeologist in Manhattan might never guess that its finest buildings were the product of a civilization in economic eclipse, marching inexorably toward global war.

For all we know, Minoan Crete was in a similar phase, even as people rebuilt upon the ruins of Mochlos. That a rebuilt home was executed in a fine architectural style tells us only that wealth must have still counted for something after Thera, and that someone whose wealth had survived returned to a relatively isolated, off-shore island (prime real estate) and employed Cretan artisans. This does not preclude the volcanic destruction of Thera having a major impact on Crete. Quite the opposite: a need to rebuild is, as a matter of sheer definition, evidence of impact.

One hundred fifty feet from the rebuilt home, pottery from a Mycenaean household has been unearthed. "It looks to me like a continuation of occupation of the neighborhood, without any real break," explains Soles. "It is as if mainland Greeks simply moved in among the inhabitants of post-Thera Mochlos, without any show of resistance from their neighbors."

Greek empire builders apparently descended from the north like an opportunistic infection, just as the bacteria which cause strep throat will settle in tissues inflamed and weakened by a flu virus (the human body's equivalent of a Thera explosion).

In the aftermath of Thera, Greece became a universal empire. A characteristic of universal empires is that they either absorb or eliminate their competitors, which invariably eliminates the need for innovation. For all their centuries of existence, Greece and Imperial Rome invented very little that was new, and eventually crumbled from within. From the moment of Greek occupation of the Minoans, both civilizations were upon the wane.

We know that mainland Greeks moved into Knossos after an episode of widespread fire destruction. The administrative writing shifted abruptly from Minoan (Linear A) to Greek (Linear B). It is difficult to imagine fleets sailing from Greece to Crete unless they expected only minimal opposition. This was not an armor-plated invasionary force equipped with guns and rockets. Bronze Age ships merely landed soldiers on beaches, if they managed to avoid being intercepted and overrun by the native fleet. Once ashore, hostile armies could easily surround the soldiers and cut off supply lines. Under normal conditions, the sea was Crete's moat and any invading force was at a severe disadvantage.

The fact remains that below Knossos' layer of burned wood, Cretans were keeping their records in Linear A. Above it, the writing is early Greek. Some archaeologists attribute the fires to anarchy and a wave of Cretan self-destruction following an earthquake and tsunamis.

Trinity College historian J. V. Luce just cannot see it that way. "Given that Crete has always been subject to earthquakes, and presumably the Cretans were used to dealing with them and rebuilding their settlements after they occurred, why would settlements be abandoned and burned only after the [1628 B.C.] disaster? What was different? And if you posit human destructions, who is doing the destroying? If Cretans, it seems rather illogical for them to further destroy their own buildings instead of putting them together again." Widespread destruction by the Greeks would also have been illogical. As the Rekhmire fresco suggests, Greek empire builders had a vested interest in maintaining Crete's commercial prosperity.

Nanno Marinatos wishes to challenge those who believe that the eruption could not have had a serious effect on the people of Crete. " 'Serious effect' need not mean a massive destruction which wiped out Crete; an economic crisis, which could have been triggered by the eruption, must have had a 'serious effect.' The evacuation of Thera alone, even before the explosion, could have disrupted Crete. The heavy burden of a great number of refugees must have produced all sorts of economic, social and political stresses.

"Professor Iakovidis has explained the fall of the Mycenaean [early Greek] world along similar lines. This economic crisis could have been produced if the Minoan fleet were destroyed (and there is no reason to assume that the *entire* fleet was safely away to sea), if the crops were burned and damaged and the animals were starved. By the way, has anybody studied volcanic effects on agriculture?

"My main point is that we do not need massive destruction of cities and fleets to explain the decline of Crete, but only an economic crisis. And we do not have to look very far for examples. During the 1930s such a thing—a global economic crisis—damned near brought down our own civilization."

"I agree entirely with Dr. Marinatos," says Professor Spyros Iakovidis. "And I would like to emphasize from the archaeological point of view a very important thing: Nobody ever talked of the civilization of Crete having been wiped out. But it seems more than probable that the results of the eruption created enough damage to bring about a hiatus, which [in turn] brought about a weakening of the Cretan economy and a political crisis."

A civilization does not need much more prodding into eclipse,

if not total extinction. In those days there were no United Nations
air lifts, no Salvation Army, no benefit rock concerts, no International Red Cross to rush in with medical supplies. The Minoans
were on their own, with Greek mainlanders waiting at the gate.
If Crete weakened and spiraled down into civil disorder—all the
more easy for Greeks to sail in and take over. It is conceivable
that, given a sufficient decline in the standard of living, the Cretans
might even have welcomed the new order.

A question that continues to nag is: What happened to Thera?
Many things, apparently.

"As far as archaeology can tell," says Iakovidis, "the town of
Akrotiri [the new name for Thera's lost city, borrowed from the
town that lies above it] was at first severely damaged by an earthquake. After this earthquake, or during this earthquake, or just
before this earthquake (in which case we must assume that it was
preceded by a few warning tremors), the inhabitants stored their
possessions as well as they could, took their valuables and left the
city. We do not know whether they were able to leave the island.
It is my opinion that in the main they were not and that somebody,
someday, will find a few thousand skeletons on some [ancient
Theran] shore. Anyway, some but not all must have escaped.
Perhaps the evacuation of the city was organized, and perhaps it
was not. I don't know. In some parts of the island it might have
been, but in other parts it obviously was not. Let me remind you
of the finds of Dr. Zahn at [the village of] Potamos—which is
very near Akrotiri, where he [discovered] several objects other
than pottery, including a very valuable [gold-] inlaid dagger—just
about the first thing someone would flee with. So it doesn't seem
to me that the withdrawal was everywhere well-organized."*

Dr. Iakovidis's Akrotiri dagger is not the only indicator of
stranded Therans and sudden death. Near Thera's north shore,
about seven miles from the excavation, miners carting away pu-

* At least one other bronze dagger, inlaid with gold, turned up in a 1988 clan-
destine dig.

mice for the Suez Canal found carbonized trees and a human skeleton. A bronze dagger with inlaid gold was also discovered, and two golden rings. I suspect that there was a second town in the north, and that some tragedy overcame its people with amazing rapidity. Whatever it was, I think it was almost simultaneous with the earthquake that ravaged the buried city. Evidence of this can be found nearer to Akrotiri, where refugees left gold in a part of the city only a few hundred yards northwest of Doumas's excavation. From all indications, the evacuation of the area so far excavated had taken time, time enough for all valuable objects to be packed away and carried off. Evidently, there was no such activity underneath Akrotiri. If any survived the quake and walked away, they did not bring their gold with them. Two different fates befell two different quarters of the same city. And something else happened seven miles away on the north shore.

It begins to seem doubtful that the earthquake alone was sufficient reason for the Therans to abandon their island paradise. Aegean people were used to earthquakes, as indicated even by the smallness of Minoan rooms and the labyrinthine arrangement of streets and buildings. The floor plan of Delta House is a maze of intersecting angles, providing the maximum structural support from a minimum of building materials. The Minoans lived in the most shock-resistant houses the Bronze Age could offer. Even so, evidence of moderate, repeated repair work on Crete suggests that Minoans had accepted as a fact of life the possibility that they'd have to rebuild every few decades.

So why should an earthquake have frightened the Therans off— especially if one of the few places to go from Thera was to earthquake-prone Crete?

My guess is that the first major earthquake was accompanied by St. Pierre–type ground surges, fiery black snakes that descended from the mountain and turned men into pillars of charcoal before they could scream or run.

The most likely reason no one returned for the dagger beneath Akrotiri is that no one who knew where it was survived the initial quake and surge.* The fact that gold was also left behind on the north shore, and that at least one body was left unburied (suggesting that other charred skeletons await discovery) hints that friends and relatives of the deceased had perished with them.

The last days of civilization on Thera probably began with the relatively minor ground shocks that are the characteristic birth cries of an awakening volcano. The people stayed, began to rebuild, and weeks, possibly months later, a major earthquake was felt throughout the island. In the excavated area, near Delta House, walls fell sideways onto ash-free ground, meaning that any Pelean death clouds that surged into or near the city were preceded by the quake, probably by a matter of seconds or minutes.

Before any significant repairs could be undertaken, a decision was made to evacuate, perhaps prompted by a ground surge striking along the northwestern edge of the city. If this display of volcanic death left any doubts about whether or not to abandon Thera, word that a settlement on the north shore (where carbonized wood and objects of gold are still occasionally recovered from the ash) had been obliterated should have settled the question, especially if, as was likely the case, the mountain continued to threaten.

And so all who remained alive on Kalliste, the island known today as Thera—fear—gathered at the docks and prepared to flee across the Aegean.

There was something both grave and reverential in that moment, something few human beings have experienced before or since, something which, perhaps, the Kallistens did not understand themselves.

They would sail away to Crete, with no thought of ever returning. And when the last of them had gone, their cities and their land would have no names.

---

* I suspect that Marinatos's squatters were people who, having heard from refugees on Crete that whole neighborhoods had been obliterated, and that fine daggers and other valuables remained unrecovered, returned to the island in search of treasure.

They couldn't hit an elephant at this dist——

——last words of General John Sedgwick
at the Battle of Spotsylvania, 1861

ON
PROBABILITY
AND
POSSIBILITY

SPRING, A.D. 1989

I can understand, now, why Arthur C. Clarke fell in love with
Sri Lanka. It reminds me very much of New Zealand, except that
the only elephants in New Zealand are at the zoo. Here they are
actually part of the morning rush-hour traffic.

I did not realize until I arrived here that the archaeological sites
around which Clarke set his novel *The Fountains of Paradise* exist
almost exactly as he described them (and the strangest thing of
all is that the fountains themselves were not discovered until after
the novel was published).

Gloria and I have climbed through the lion's jaws to the ancient
brick fortress atop Demon Rock, and explored the giant cisterns
there—which, judging from the stairways descending into them,
and the artificial ledges (false beaches), must have served also as
palace swimming pools. The sacred mountain also exists, rising
1.4 miles above the tropical forests. Two stairways—the world's
longest—lead to a mountaintop temple. During each full moon,

hundreds of torch-bearing pilgrims climb the stairs, starting at midnight and arriving at dawn to witness the mountain's peak cast as a triangular shadow in the western sky.

I gave a lecture at the Arthur C. Clarke Center yesterday: An Introduction to Antimatter Propulsion and Interstellar Flight. They're building a new satellite station there—which seems to be keeping the students quite busy. After the lecture, we went across the river to another of Arthur's favorite haunts: the Institute of Integral Education. Its founder, Dr. Mervyn Fernando, had invited us for afternoon tea. He is a Jesuit priest with a keen interest in archaeology. And what archaeological story is more interesting than the one emerging from Thera?

Clarke is the man who invented the communications satellite, and one of the world's most extraordinary dreamers. Long before anyone had gone to the moon, even before the first satellite had been orbited, he had written "The Star," a story about a Jesuit priest's voyage to another solar system, where an extinct civilization was being excavated. The civilization, a remarkably refined one, had died under the glare of a supernova—which turned out to be the star of Bethlehem. I couldn't help remarking on the parallels to Thera and the Nile delta ash.

"Except for one thing," said Clarke. "Atlantis and the plagues of Egypt are nowhere near as dramatic as a supernova."

"But the paradox was the same," I pointed out. "If God wanted to strike fear into the pharaoh, why did he need Thera? Why was it necessary to destroy innocents hundreds of miles away?"

"True. But in my story the protagonist was a Jesuit priest who found his faith shaken by the paradox. It caused him great mental anguish. He began to wonder if the crucifix on his wall was no more than an empty symbol. Now, Father Fernando tells me that I underestimated the Jesuits."

Father Fernando gave a pleasant smile. "Yes. It is the sort of news we would gladly take to the Vatican."

"A Jesuit friend of mine in New York has told me the same

thing," I said. "But I do not quite understand. He referred to something called 'happy fault.' "

"*Felix culpa*, the exalted," explained Father Fernando. "During Holy Week there is a chant, a beautiful intonation: 'Oh, happy fault, oh, necessary sin of Adam, which gained for us so great a redeemer.' It's the paradox of the cross, wherein redemption comes from the destruction of an innocent."

"So the Thera paradox, from one perspective, can be viewed as a paradox of redemption. The destruction of Thera is what frightened the pharaoh—Tuthmosis III, apparently—into letting Moses and his people go."

"According to that one possible viewpoint, Charles, what seems to be the destruction of life in the Minoan world is what brings life somewhere else, in this case to the children of Israel."

"You'd have to be a jolly optimist to adopt that viewpoint."

"It's a message that recurs throughout the Bible: Cain turns destructive and evil, becomes the legendary first murderer, but God puts a mark upon his head, so no one will kill him, that he might survive to find redemption."

"But there were so many other ways of frightening the pharaoh. If we are to believe that this was God's will, would a just God require that for an act of good in Egypt there had to be an act of evil in the Aegean? Does a just universe absolutely require the destruction of Thera?"

There was silence for a moment. Then for another. As the sun went behind a cloud, a housekeeper brought a tray of tea onto the open-air veranda; and a cool breeze came up, and across the river great shades of light and shadow seemed to play over the rain forest and the satellite tower. Men think they have power with their satellites and their plutonium fists, but we forget too easily that the earth has power of its own.

When Captain Jacques-Yves Cousteau visited Thera in 1976, he invited Doumas aboard the diving saucer, to seek the bottom of the flooded crater. But Doumas knew that nothing archaeological could have survived in the explosion center, nothing to interest him in going beneath a thousand feet of water and hovering in the mouth of a live volcano, so he declined the offer.

Cousteau never did find the bottom of the crater. It dropped down deeper than the saucer could go—at least a third of a mile. And yet the hole had been filling since the eruption; a new mountain had sprouted in its center, erasing most of its features. Whatever must the original crater have been like? In that autumn of 1628 B.C., in the immediate aftermath of the explosion, it had to be at least a mile deep. And as the loudest sound ever heard by human ears spread out over the earth, and the walls of water hovering around the hole edged in and eventually swallowed the crater, there must have been, for a while, waterfalls unlike anything ever seen on earth. The *Queen Elizabeth II* going over the edge would have looked smaller than a war canoe going over Niagara. One might never have noticed it.

Finally, Clarke broke the silence.

"According to one view of the Creation, once God set the laws of the cosmos into motion—atomic mass, atomic weight, the bonding habits of carbon and hydrogen—no one could tamper with them again, not even God."

"That limits God today to the role of innocent bystander," objected Father Fernando. "I don't think so. I suspect that, despite the depressing difficulties of the present, mankind is being called to a great and grand destiny."

"Perhaps intelligent species everywhere have grand destinies," I said. "Life itself seems to have been inevitable from the beginning. There was no choice in the matter. We are the result of some of the most likely chemical reactions undergone by some of the most abundant elements in this part of the universe. Perhaps it is with the emergence of intelligence that free will finally enters the universe. Long before the earth existed, the position and motion of every particle in the cosmos probably dictated that in 1628 B.C., Thera was going to explode, and on April 14, 1912, a seventy-mile-long ice field was going to form on the North Atlantic. Not even God could have prevented it. But unless all the events of past, present and future, including the decisions we make during life, are laid out across time and space like frames on a movie reel, men had the free will to heed the warning signs. In the case of the *Titanic*, for example, they did not have to think they were more powerful than nature and ride full steam ahead through the dark into an ice field they knew was there.

"If you believe in Chaos Theory," said Arthur, "this should be impossible—nothing can be predicted in detail. So God has to create the universe just to see what will happen. Even God does not know. But you believe Thera would have exploded in 1628 B.C., and the ice would have formed in A.D. 1912 no matter what man or God tried to do?"

"Absolutely. But Thera was different from the *Titanic*, wasn't it? Whether or not the Therans heeded the warning signs and evacuated the island, they were still going to suffer. The people in charge of the *Titanic* had a choice. The Minoans did not. When you come right down to it, even the pharaoh had no choice. If he had told Moses, 'Go! Get out of here. And take your people with you,' the volcano would have exploded anyway, and Egypt would still have suffered under the ash cloud."

"There's a mystery here," said Father Fernando, "but let me try a few ideas. If you view history as a tapestry, an interaction of physical events and moral behavior, does four thousand years of widespread general violence have any particular significance? Was the violence at Thera just accidental, with no particular connection to the present historical moment? Was it just one of those downswings in the ups and downs of man's historical journey, to be followed by the next upswing? Or was it just another meaningless episode in the total meaninglessness of man's existence? Or did it play some role in the progressive evolution of man into another stage of his future?

"Consider this: man is the only being that is conscious that he is conscious. Consciousness doubles back on itself, seeks its origins. Going downward past Thera and man we see decreasing

degrees of consciousness with decreasing degrees of complexity, down past our furry ancestors, and our finned ones before them, right down to the level of inert matter—molecules and atoms.*
Seen against this wide scope, the Thera explosion, the Crusades, the Nazis, all of these things were temporary outbursts of evil, interruptions in the flow—"

"But—"

"I know what you're going to say, Charles. Wars are man-made outbursts—'choice evil.' Thera was not—it was a 'physical evil.' You believe that man had no choice about Thera? But even in the Aegean there must have been some degree of choice. Did you not tell us that there was evidence of conquest on Crete following the eruption? Did man not respond to nature's violence with violence of his own?"

He was right, of course. And if one looked at the relationship between man and earth today, it was possible to believe that civilization, though outwardly more advanced, had taken a giant leap backward since Minoan times. Nature was threatening new violence, this time in direct response to man-made outbursts (what Father Fernando would call choice evil) against the earth itself.

When Captain Cousteau descended into Thera Lagoon, he found no hints of artifacts from an ancient civilization, but there were artifacts aplenty from our own time: beer cans, soda bottles, the odd sneaker. When God met man in Eden, Cousteau had reflected to Doumas, He commanded us to replenish the earth and subdue it.

"But in order to do that," Cousteau said, "we have to remember that the earth has a life of its own. That's when people get in trouble: when they have no respect for the earth, for its power, its majesty. We need it to survive, but it doesn't need us. The planet existed four and a half billion years without us, and if we're not very careful, we may be as extinct as the brontosaurs one day.

* For another view on this subject, Stephen Jay Gould's essay, "Tires and Sandals," in the April 1989 issue of *Natural History*, is must reading.

"When you look at the buried city, or visit the frescoes in the Athens museum, you see the poised serenity of a Golden Age in art, the innocence of a civilization that believed it might live forever. . . . On the crater rim, in Fira, the artists no longer paint frescoes. Yet life goes on. Each day is a reminder that man has always lived on the edge of extinction by earthquake or plague, by famine or his own folly. Here, each man knows it is best to turn one's own field, and be thrifty with water, keep peace with neighbors, pray to whatever gods are handy. And if your wings are made of wax, do not fly too close to the sun."

Plato's Atlantis is a cautionary tale about a civilization that vanished unexpectedly, in a day, at the height of its glory. People do not and never have liked to believe that a whole civilization can be destroyed without reason, by a random act of nature. People seek order in the universe. In an orderly universe, islands and cities do not sink beneath the waves unless their inhabitants were bad at something that other people are good at. And so, after the dust had settled, after the first ships had explored the smoking crater, the story began passing by word of mouth from one generation to another and—people being people—some kind of lesson was sought from the whole damned mess. The victims became villains because they vanished, vanished because they were villains. The legend has come down through the ages with the moral that the Atlanteans grew too proud, and were punished for their arrogance. The excavations at Thera tell a different story. An arrogant people would not have heeded the volcano's warnings. They would have stayed behind, clinging to their wealth, believing fully in the solidity of their great city.

In more pessimistic moments, it has been easy for me to imagine that if there are sentient beings amidst the stars, they may carry tales one day of a people who were on the verge of bridging interstellar space, yet in a geological nanosecond disappeared utterly and without descendants. Sooner or later, they'd come here, to sink shafts into the debris field over Manhattan, to visit *Apollo*

11 on the Sea of Tranquillity, in an effort to determine whether we really existed or if the legend was merely that—a cautionary tale.

"The first thing that must be asked about future man," Charles Darwin said, "is whether he will be alive, and will know how to keep alive, and not whether it is a good thing that he should be alive."

"The first thing I'd like to ask about future man," Father Fernando resumed, "is whether the evolutionary process stops with our present civilization, or are there further stages ahead? If you go back to the origins of civilization, if you are really paying attention, you can get bearings and sights to peer into the future. As the diversity of life on earth increased, some organisms became increasingly complex. Increasing complexity brought increasing consciousness, and now one conscious organism has spread from pole to pole, over the oceans and under them, enclosing the earth in a single thinking envelope. And the envelope is becoming more and more unified. We see this happening before our very eyes."

Father Fernando glanced out toward the river. On the shore, a breeze stirred broad-leaved evergreens. They rustled secretly. "Across the river, at the Arthur C. Clarke Center, that satellite tower rising out of the forest is an example of the links and bonds that keep forming every moment between person and person, culture and culture, continent and continent. . . .

"We fail to realize, I believe, how new this experience is for mankind. For almost the whole of man's million or so years of existence as *Homo sapiens*, he has lived in small tribal groups in scattered isolation on the face of the earth.* The fantastic devel-

* Though some may be surprised to hear a priest making references to human evolution, the popular notion that there is a rift between science and Christianity, and that all churches must therefore reject the idea that continents and species have been evolving for millions and billions of years, is a myth. Indeed, in 1982, one group of distinguished scientists issued a formal statement on the matter: *We are convinced that masses of evidence render the application of the concept of evolution to man and the other primates beyond serious dispute.* Few people seemed to take notice, though the statement should have settled the matter, for even the most vocal proponent of "scientific creationism" had to admit that the

opment of rapid transport and communication in this century has bundled us up in a web. The web is to mankind what the nervous system is to the body of the individual. We are becoming a single grain of thought on the cosmic scale.

"It might have happened a little sooner if Thera had not exploded; but sooner or later it would have happened, no matter what the natural or man-made obstacles."

The way Father Fernando viewed civilization, five billion people, seemingly mindless of what was actually happening, were creating a world mind. The planet was acquiring a nervous system. Who knew what shape it would eventually take? It occurred to him that the *Voyager* and *Pioneer* space probes provided a clue. Already, at a fetal stage, earth had begun detaching pieces of itself, flinging motile ganglia at the stars. Whether or not a supreme being had a hand in the Creation (a requirement that, as Arthur later pointed out, dictated a Creator more complex than the universe it has created, and thus more than doubled the size of the original problem), once the first stars had begun exhaling something as versatile as the carbon atom, its progression, in the presence of a warm, moist planet, in the direction of methane, porphyrins, membranes, respiration, variation, speciation and consciousness was, sooner or later, bound to spawn a world mind. A volcano might have put the process back a few dozen centuries; global violence and a failure to cherish the earth may yet destroy it, "but the web is unavoidable," said the Jesuit. "Man is now being called upon to go freely and consciously into his evolutionary vocation—which is a vocation of unity and convergence.

"Given this premise, the violence of Mycenaean Greeks against the volcano-weakened Minoans, and the violence that divides much of the world today, even the violence we inflict upon nature, are man's refusal to respond positively to his evolutionary destiny.

Vatican did speak with a certain authority on matters of faith. With the pope's blessing, the Pontifical Academy of Sciences announced that evolution was "beyond serious dispute."

It is up to us as free and conscious beings to make or mar our future.''

"Arthur?''

"Well, there's nothing much to say,'' Clarke said slowly, "nothing much that I haven't said already in half a dozen books. Most people have trouble just with the *name* God, never mind interpreting her will. The trouble is that the name has never meant quite the same thing to any two people. That's how you get new religions cropping up all the time, much like the branching process from which new species arise. Judaism branches off from some ancient Egyptian or Babylonian sect, Christianity branches from Judaism, incorporating bits of both Egyptian and Roman mythology into the Holy Roman Catholic Church. Then the Muslims branch off from the Christians, and both religions continue branching, as does Judaism, into multiple new sects running the spectrum from snake handlers to America's TV evangelists, each claiming to possess the One and Only Truth, half of them going to war with each other—and that's just the story of the Mediterranean gods.

"Many people seem to be turning to the notion of a Creator who set the universe in motion and might or might not have had anything to do with it ever since. It's almost a new religion in its own right, and I think this arises from a growing perception of randomness in the world. Volcanoes explode and kill innocents. Planeloads of children are lost. And then there are anomalies like Frank 'Lucks' Towers, the sailor who survived the *Titanic*, the *Empress of Ireland*, and the *Lusitania*—after which he decided to take up farming. But in a world of five billion human beings, there are bound to be a couple of million who seem to get more than their share of good luck or bad luck. Statistically, such anomalies are inevitable, given enough people: five billion drawings of the cards. It's all within the odds. That's what I think this Atlantis legend is all about: just one of those unlucky things. The Minoans had the good fortune of being located on a group of islands whose very geography encouraged people to compete with each other, yet prevented wholesale takeover and stagnation. It was precisely the environmental setting most likely to spawn an advanced civilization, yet the very geology that created the setting also provided for its destruction. Where was divine intervention?

"When you really stop and look around, even if any of us sitting here today were to examine every moment of a given week of our lives in detail, I think we would see that bad things happen as often as good. We tend to remember only the extraordinary events, such as the odd coincidences; but we forget that almost every event is an odd coincidence. The asteroid that misses the earth is on a course every bit as improbable as the one that strikes it. We cheer a royal flush, but every hand is equally improbable as a royal flush. There's nothing extraordinary about it at all.

"One day, when computers are sufficiently advanced to permit scrutiny of events all over the world in vivid detail over the course of several years, I suspect we will learn that the pattern of bad and good events is utterly randomized, that the world simply obeys the laws of mathematical probability, that we can discern no signs of supernatural intervention, either for good or evil.

"But you must remember one thing, Charles."

"What is that?"

"To question everything. Promise me you'll do that. Promise me you'll not take on faith a word Father Fernando or I have said—merely because it happened to come from our mouths. Good and evil, origins and extinctions, the loss of the Minoans . . . questions in every direction. And let me tell you something: civilization is still very young, and the universe is a mysterious, magical place in which to be growing up—and sometimes I still wonder about God."

AFTERWORD:
WHERE ARE THEY TODAY?

Robert Ballard continues to explore the beds of the Atlantic and Pacific oceans, particularly the forty-thousand-mile-long volcanic mountain range that, somewhat like the seam on a baseball, completely girdles the planet.

Auguste Ciparis was discovered in his prison cell three days after the eruption of Mount Pelée. His sentence was suspended and he became famous for being the sole survivor of St. Pierre. His claim to fame was the invention of P. T. Barnum, who knew very well that Ciparis was not the only survivor; but the story turned the young man into a very valuable asset for a living exhibit Mr. Barnum had in mind. He offered Ciparis a job with his circus, and the newly free man began living behind bars in a traveling replica of his cell. He was still there when he died in 1929 at the age of forty-five.

Arthur C. Clarke lives in Sri Lanka, where he supports educational institutions, advises scientists all over the world, writes several books a year and moves mountains.

Fernand Clerc survived the May 8, 1902, eruption of Mount Pelée, and remained with his family on the island of Martinique until his death in 1921.

Captain Jacques-Yves Cousteau, though into his eighties, still leads expeditions into the world's waterways, and fights for their preservation.

Christos Doumas continues as director of the Thera excavation, though he has been involved in a bitter dispute over the lack of funding even to protect the site, much less excavate it further. As of this writing, the future of the lost city appears uncertain.

Father Mervyn Fernando continues as director of Subodhi, the Institute of Integral Education, Sri Lanka. During the civil disturbances of the 1980s, while universities remained closed, Subodhi and the Arthur C. Clarke Center were among the very few educational institutions allowed to stay open.

K. T. Frost first suggested in 1909 that the legend of Atlantis was based upon a genuine but garbled memory of the newly discovered Minoan civilization, but his ideas were not taken seriously until Spyridon Marinatos began looking into the matter just before the onset of World War II. Frost did not live to see this. He was killed in action during World War I.

Queen Hatshepsut appears to have ruled Egypt during the decades preceding the explosion of Thera (1628 B.C.). Her stepson, Tuthmosis III, desecrated a portion of her sarcophagus, chiseling away the golden sculpture of her face to prevent her resurrection from the afterlife. The sarcophagus was discovered early in this century and eventually made its way to a display in the Cairo Museum. There is a persistent rumor in Egypt that Hatshepsut left behind an evil curse on anyone who despoiled her grave or attempted to remove its contents from the shores of the Nile. Tuthmosis III was apparently the first to feel the curse, in the form of Theran ash, seven plagues, an exodus of Hebrew slaves, and the annihilation of the fighting force that pursued them. In 1910, the British Egyptologist Douglass Murray is said to have been among the next to feel the curse. An American treasure hunter approached him in Cairo, offering for sale one of several portions of Hatshepsut's multilayered mummy case. The American died before he could cash Murray's check. Three days later, Murray's gun exploded, blowing off most of his right hand. What remained turned gangrenous, requiring amputation of the entire arm at the elbow. En route to England with the sarcophagus, he received word via wireless telegraph that two of his closest friends and two of his servants had died suddenly. Upon arrival in England, he began to feel superstitious, and left Hatshepsut's case in the house of a girlfriend who had taken a fancy to it. The girlfriend soon came down with a mysterious wasting disease. Then her mother died suddenly. Her lawyer delivered the sarcophagus back to Murray, who promptly unloaded it on the British Museum. The British Museum already had more sarcophagi than it needed. Not so, the American Museum of Natural History in New York; so a trade was arranged for Montana dinosaur bones. Before the deal was complete, the British Museum's director of Egyptology and his photographer were dead. Queen Hatshepsut was beginning to lose her charm. The curators loaded the sarcophagus into a crate and saw it lowered into the hold of a ship. And on April 10, 1912, the ship sailed—Southampton to New York. Hatshepsut departed for America on the *Titanic.*

Gaston Landes, the geologist who assured Governor Louis Mouttet that Mount Pelée was safe, dove to the bottom of a pond when he saw the death cloud snaking toward him. The water shielded him from the glowing dust and scalding air . . . until he came up for air. When Fernand Clerc and his party entered the ruins in search of survivors, they found him flopping helplessly at the edge of the pond. His tongue had been seared from his mouth, his eyes from their sockets, and the rest of his face was a massive blister. One of the searchers asked him if he had seen her husband. The others gently eased her away, and moments later Landes was dead.

Léon Compère-Léandre, who had survived the blast from Mount Pelée while napping in a wood cellar, was found running along Le Trace, away from the dead city. Rescuers took him to the town of Fort-de-France, where hospital workers promptly labeled him a madman. Apparently, this diagnosis did not disqualify him from appointment to Martinique's police force, for two days later he was given a gun and sent back to St. Pierre to guard the ruins against looters. On May 20, 1902, after barely more than a week of service, he abandoned his post and began walking in the direction of Fort-de-France. Behind him he heard "a great awhump!" When he looked back, he saw another death cloud descending upon the city, washing away whatever had been spared by the May 8 eruption. Three months later, on the morning of August 30, Mount Pelée exploded again, this time sending a cloud crashing and burning through the village of Morne Rouge, where Léon had just taken up residence. He emerged as one of its few survivors, yet continued to live on the island of Martinique until his death from a fatal fall in 1936 (he slipped in a bathtub and broke his neck).

Walter Lord, the historian who happened to converge on Thera at the same time as Spyridon Marinatos, Arthur C. Clarke and Wernher von Braun, continues, along with Isaac Asimov and the New York Public Library, to be one of America's natural resources.

Love Canal, site of the world's most infamous toxic waste disaster, continues to crumble in silence. It is a very modern-appearing ghost town, and because it looks so ordinary and familiar, yet at the same time distorted and empty, it is one of the strangest, most chilling places on earth. Even more chilling are claims by government "experts" that portions of the town are safe for human habitation and should be resettled. Don't believe it. When those same experts visit the site, they reside in an atmosphere-controlled building with its own air supply, and drink only bottled water.

Nanno Marinatos, daughter of the discoverer of Thera's lost city, became an archaeologist herself and now continues her father's work. She notes that she often arrives at different conclusions from him, and has even disproved one or two of his theories. The Minoan-Atlantis connection is still around—though sometimes she wonders about its validity. Since scientific advance depends upon questions and doubts, there can be little doubt that Spyridon Marinatos (though gnashing his teeth a little) would have been proud of her.

Governor Louis Mouttet, who had ignored the warning signs of an awakening volcano and even posted troops to prevent people from leaving St. Pierre until the election was over, died with his constituents in the eruption of May 8, 1902. His body was never found.

St. Pierre was rebuilt exactly where it fell, in the shadow of Mount Pelée, which will surely erupt again. Tourist brochures explain that Pelée is an extinct and safe volcano, and tour guides and cabbies often deny that a volcano exists on the island.

Edith Russell, whose memoirs recounts the peculiar behavioral response that bears her name, survived her voyage aboard the *Titanic* quite by accident. She was, at the time, a fashion reporter, but seems to have subsequently become addicted to action and adventure. When World War I broke out, she was living on the front lines with K. T. Frost and other defenders of the British Empire. There, in the trenches, she became the world's first woman war correspondent. Even when she was not looking for action, disaster seems to have had a way of following her. By the time she died, on April 4, 1975, she'd survived bombings, car accidents, fires, floods, tornados and at least one more shipwreck. Boarding an airplane, she decided, would be daring God. So she went everywhere by steamship or car and, at the age of ninety-

eight, was able to say she'd had every disaster except a plane crash,
a husband and bubonic plague.

Ellery Scott, first officer of the ill-fated *Roraima*, continued to sail, even
through the action of World War I. He retired from the sea in 1923,
and whatever happened to him beyond that point remains a mystery.
The *Roraima* itself lies submerged and rusting in the harbor of St.
Pierre.

Daniel Stanley, a discoverer of the Theran ash layer in the Nile Delta,
continues to study the deposit, with particular emphasis on reconstructing the environment at the time of the ashfall.

Harry Truman was probably so thoroughly vaporized by Mount St.
Helens that he literally became part of the earth's atmosphere. If you
are old enough to be reading this book, chances are you have already
breathed a few atoms of him.

Tuthmosis III, the apparent pharaoh of both the Thera explosion and
the Exodus, and the first Egyptian conqueror of the eastern Mediterranean world, died of natural causes at advanced old age. Despite a
labyrinth of false walls and death traps, robbers entered his tomb by
1000 B.C. and removed almost everything of value, including the
pharaoh's golden death mask. Gemstones and sheets of gold were
stripped away from his wooden sarcophagus, and little more than the
mummy itself and objects of wood or alabaster remained. The same
fate befell every pharaoh except Tutankhamen, who still sleeps in his
golden casket beneath the Valley of the Kings. Elsewhere in the valley,
thirteen desecrated royal mummies, including Tuthmosis III, were
eventually sheltered in the tomb of Amenophis II. This tomb, too,
was broken into. The greater part of whatever art treasures remained
were spirited away and dismembered for their gold and jewels. Most
of the funeral equipment—delicate alabaster containers filled with oil
and perfumes—had been drained of their contents and smashed against
the walls. And then, for nearly three thousand years, the tomb was
miraculously forgotten. When archaeologists entered it in 1898, they
found the mummies intact. Much had been taken, it was true, but
the shelter had escaped the total destruction seen in other tombs. The
body of Tuthmosis III himself still lay within its own (albeit plundered)
sarcophagus, seeming to have been spared the final indignity of being
stripped completely and burned for fuel; but within a year modern
tomb robbers, doubtless with the cooperation of the guards, removed
almost all that had survived. The pharaoh's funerary wrappings were
cut and torn, and the body within was searched and stripped of rings

and other jewelry. What remained of the pharaoh was removed to the Cairo Museum where, today, a visitor can look upon the face that looked upon Moses.

From the desert sands surrounding Cairo, archaeologists of the nineteenth century A.D. recovered an obelisk inscribed with the name Tuthmosis III. Erected around 1620 B.C., the pharaoh intended the huge limestone needle to survive into our time and make his name immortal. It now stands in Central Park, New York, where, as it slowly dissolves under layers of moss, acid rain and pigeon droppings, tourists who pose for pictures beneath it may be heard to ask, "Why do they call that pillar Cleopatra's Needle?"

Wernher von Braun, after leading the drive to put America on the moon, was discarded along with most of the hardware he created. Incredibly, most of the actual instructions for the construction of Apollo moonships now lie moldering in such places as Long Island's Oceanside landfill. Toward the end of his life, von Braun sold helicopters. He was among the lucky ones. Some of his colleagues wound up selling hot dogs at the entrances to Central Park, in the very shadow of Cleopatra's Needle. The disposal of such talent explains largely why no one in America knows how to build a Saturn rocket today, and why the *Challenger* exploded (no Apollo veteran would ever have cleared her for launch). Those closest to von Braun believe the cancer that eventually killed him was related to disappointment and depression that became too large. He died on June 16, 1977—died in frustration and obscurity only eight years after the triumph of *Apollo 11* on the Sea of Tranquillity.

Voyager 2, along with three other robot spacecraft now hurtling out of the solar system, is among the most timeless objects ever built by the hand of man. The same erosive forces that are slowly tearing down the Rockies and the Himalayas have already reduced the Pyramids to ghost images of their former glory. On the airless, erosionless moon, remnants of Apollo will still be shiny and recognizable long after Everest is gone; but even these will vanish when the sun swells to red giantism and engulfs everything out to the orbit of Mars. Yet thousands of light years from home, *Voyager 2* will still speak for earth, carrying with it a computer disk engraved with pictures and voices from our time—even samples of our art and our music. Ten billion years after the sun reduces the earth to slag, our spacefaring robots may be the only evidence that human beings existed.

SELECTED
BIBLIOGRAPHY

Attenborough, D. *The First Eden: The Mediterranean World and Man.* Boston: Little, Brown & Company, 1987.

Badeau, J., et al. *The Genius of Arab Civilization.* Oxford: Phaidon Press, 1978.

Baines, J., and J. Malek. *Atlas of Ancient Egypt.* Oxford: Phaidon Press, 1980.

Benford, G. *Timescape.* New York: Simon and Schuster, 1980. (Fiction: "If the reader emerges with the conviction that time represents a fundamental riddle in modern physics, this book will have served its purpose." —G. Benford)

Brownlee, S. "Explorers of Dark Frontiers." *Discover,* February, 1986.

Burke, J. *The Day the Universe Changed.* Boston: Little, Brown & Co., 1985.

Cadogan, G. "Archaeology: Unsteady Date of a Big Bang." *Nature,* 238 (1987), p. 473.

———. "Volcanic Glass Shards in Late Minoan I Crete." *Antiquity,* 46 (1972), pp. 310–313.

Carter, H. *The Tomb of Tutankhamen.* London: Century Hutchinson Ltd., 1983 (originally published in installments between 1923 and 1933).

Clarke, A. C., "The Star" (short fiction) in *The Other Side of the Sky*, New York: Signet, 1959.

Clarke, A. C., S. Hawking and C. Sagan. *God, the Universe and Everything*. London: BBC Television, 1990 (television—educational program of the Must Viewing kind).

Clerc, F., et al. "Correspondence Relating to the Volcanic Eruptions on Martinique." *His Majesty's Accounts and Papers*, Vol. LXVI, London, September 1902.

———. "Further Correspondence Relating to the Volcanic Eruptions in St. Vincent and Martinique in 1902 and 1903." His Majesty's Stationery Office, London, 1903.

Cook, J. M. *The Greeks in Ionia and the East*. London: Thames and Hudson, 1962.

Cott, J. *The Search for Omm Sety*. New York: Warner, 1987.

Cunliffe, B. *Rome and Her Empire*. London: The Bodley Head, 1978.

The Daily News, Longview, Washington, and *The Journal American*, Bellevue, Washington, combined staffs. *Volcano: The Eruption of Mount St. Helens*. Seattle: Madrona Publishers, 1980.

Deiss, J. J. *Herculaneum: Italy's Buried Treasure*. New York: Harper and Row, 1985.

Doumas, C. "The Minoan Eruption of the Santorini [Thera] Volcano." *Antiquity*, 48 (1982), pp. 110–115.

———. "The Minoan Thalassocracy and the Cyclades." *Archaeologischer Anzeiger* (1982), pp. 182–210.

———. *Thera: Pompeii of the Ancient Aegean*. London: Thames and Hudson, 1983.

Doumas, C., et al. *The Excavations of Akrotiri, Thera:* Vol. I, *Delta House;* Vol. II, *West House*. Athens: in preparation.

———. *Thera and the Aegean World*, Vols. 1 and 2. Athens: G. Tsiveriotis, Ltd., 1978, 1980. (Note: almost all of Doumas's publications postdate the death of Spyridon Marinatos.)

———. Vol 3. Athens: G. Tsiveriotis, Ltd., 1990 (proceedings of the 1989 Thera conference), in press.

Erbstosser, M. *The Crusades*. Devon, England: David & Charles, 1978.

Federman, A., et al. "Volume and Extent of the Minoan Tephra from Santorini [Thera] Volcano: New Evidence from Deep-Sea Sediment Cores." *Nature*, 271 (1978), pp. 122–126.

Findley, R. "Mount St. Helens." *National Geographic*, 159 (1981), pp. 3–65.

Francheteau, J., R. Hekinian and R. D. Ballard. "Morphology and Evolution of Hydrothermal Deposits at the Axis of the East Pacific Rise." *Oceanologica Acta*, Vol. 8 (1985).

Frost, K. T. "The *Critias* and Minoan Crete." *Journal of Hellenic Studies*, 33 (1908), pp. 189–206.

Gould, S. J. *Ever since Darwin*. New York: W. W. Norton, 1977.

———. *The Flamingo's Smile*. New York: W. W. Norton, 1985.

———. *Hen's Teeth and Horse's Toes*. New York: W. W. Norton, 1983.

———. *The Panda's Thumb*. New York: W. W. Norton, 1980.

———. *Wonderful Life*. New York: W. W. Norton, 1989.

Hagg, R., and N. Marinatos. *The Minoan Thalassocracy: Myth and Reality*. Stockholm, 1984.

Hammer, C. U., et al. "Dating of the Santorini [Thera] Eruption." *Nature*, 332 (1988), pp. 401–402.

———. "The Minoan Eruption of Santorini [Thera] in Greece Dated to 1645 B.C.?" *Nature*, 238 (1987), pp. 517–519.

Heilprin, A. *The Eruption of Pelée*. Philadelphia: J. B. Lippincott Co., 1908.

Hekinian, R. "Undersea Volcanoes." *Scientific American*, July 1984.

Hood, S. *The Minoans*. London: Thames and Hudson, 1971.

Hsu, K. *The Mediterranean Was a Desert*. Princeton, N.J.: Princeton University Press, 1983.

Hughes, M. K. "Ice Layer Dating of Eruption at Santorini [Thera]." *Nature*, 335 (1988), pp. 211–212.

Judge, J. "Minoans and Mycenaeans, Sea Kings of the Aegean." *National Geographic*, 153 (1978), pp. 142–185.

Klingender, F. *Animals in Art and Thought to the End of the Middle Ages*. London: Routledge & Kegan Paul, 1971.

Lloyd, G.E.R. *Early Greek Science: Thales to Aristotle*. London: Chatto & Windus, 1970.

Luce, J. V. *Lost Atlantis: New Light on an Old Legend*. New York: McGraw-Hill Book Co., 1969.

———. "More Thoughts About Thera." *Greece and Rome*, 19 (1972), pp. 35–46.

Marinatos, N. *Art and Religion in Thera: Reconstructing a Bronze Age Society*. Athens: Andromedas I, 1984.

Marinatos, S. *Excavations at Thera, 1968–1974*. Athens: The Athens Museum, 1975.

———. "On the Chronological Sequence of Thera's Catastrophes." *Acta* (1971), pp. 403–406.

————. *Some Words About the Legend of Atlantis*. Athens: The Athens Museum, 1969.

————. "Thera: Key to the Riddle of Minos." *National Geographic*, 141 (1972), pp. 702–726.

————. "The Volcanic Destruction of Minoan Crete." *Antiquity*, 13 (1939), pp. 425–439.

McPhee, J. *Basin and Range*. New York: Farrar, Straus and Giroux, 1980.

Mellaart, J. *Catal Huyuk*. London: Thames and Hudson, 1967.

Meyers, R. J. "An Instance of the Pitfalls Prevalent in Graveyard Research." *Biometrics*, 19 (1963), pp. 643–650.

Michael, H., and G. Weinstein. "Radiocarbon Dates from the Destruction Level of Akrotiri, Thera." Philadelphia: *Temple University Aegean Symposium*, 1977, pp. 27–30.

"Miscellaneous Observations on the Volcanic Eruptions at the Islands of Java and Sumbawa, with a Particular Account of the Mud Volcano at Grobogar." *Quarterly Journal of Science and the Arts*, 1 (1816), pp. 245–258.

Money, J. H. "The Destruction of Akrotiri." *Antiquity*, 47 (1973), pp. 50–53.

Page, D. L. "The Santorini [Thera] Volcano and the Destruction/Desolation of Minoan Crete." *The Society for the Promotion of Hellenic Studies, Supplementary Paper 12*. London: International University Booksellers, 1970.

Pang, K. D. "The Legacies of Eruption." *The Sciences*, 31 (1991), pp. 30–35.

Pellegrino, C. R. *Her Name, Titanic*. New York: McGraw-Hill Book Co., 1988.

————. "The Trouble with Nemesis," *Evolutionary Theory*, Vol. 7 (December 1985), pp. 219–221.

Pellegrino, C. R., and J. A. Stoff. *Darwin's Universe: Origins and Crises in the History of Life*. (Originally published: New York: Van Nostrand Reinhold, 1983), Blue Ridge Summit, Pa.: TAB Books, 1986.

Pichler, H., and W. Schiering. "The Thera Eruption and Late Minoan I B Destructions on Crete." *Nature*, 267 (1977), pp. 819–822.

Prichard, J. B. *Ancient Near Eastern Texts*. Princeton, N.J.: Princeton University Press, 1955.

Rapp, G., et al. "Pumice from Thera (Santorini) Identified from Greek Mainland Archaeological Excavation." *Science*, 179 (1973), pp. 471–473.

Siebold, L. "A Visit to the Dead City." *Colliers Illustrated Weekly*, June 14, 1902.

Stanley, D., and H. Cheng. "Cores of Santorini [Thera] Ash Layer in the Nile Delta." *Science*, 240 (1987), pp. 497–500.

———. "Volcanic Shards from Santorini [Thera] in Minoan Ash in the Nile Delta, Egypt." *Nature*, 320 (1986), pp. 733–735.

Sullivan, D. G. "The Discovery of Santorini [Thera] Minoan Tephra in Western Turkey." *Nature*, 333 (1988), pp. 552–554.

Symons, G. J. *The Eruption of Krakatoa and Subsequent Phenomena*. London: Report of the Krakatoa Committee of the Royal Society, 1888.

Thomas, G., and M. Morgan Witts. *The Day Their World Ended: The Destruction of St. Pierre and All Its 30,000 Inhabitants*. New York: Stein and Day, 1969.

Vitaliano, D. B., and C. J. Vitaliano. "Volcanic Tephra on Crete." *American Journal of Archaeology*, 78 (1974), pp. 19–24.

Warren, P. *The Aegean Civilization*. Oxford: Elsevier, 1975.

Wilson, A. W. *Neptune's Revenge*. New York: Bantam Books, 1985.

INDEX

A

acid rain, 77, 99, 179, 235, 236, 264

Aegean Sea, 7, 13, 15, 16, 17, 39, 117, 140, 200, 233

Africa, 16, 56, 102, 115, 123, 126, 127, 139, 146, 274
  continental drift of, 37, 128, 130, 131–32, 133, 135, 136, 137

Akhenaton (Amenophis IV), 241

Akrotiri, 20, 147, 157, 167, 169, 191, 192, 219, 221, 261, 262, 280, 281–82

Alaska, 132, 133

Alexandria, 90, 108–9, 113, 273
  library at, 107, 111, 198

algae, 97, 130, 137, 139, 264

Alkaid, 110, 117–18, 129, 134

Allende meteoroid, 36

Alvin, 33–34, 36, 58

amber, 164–65

Amenophis II, 44, 301

Amenophis IV (Akhenaton), 241

Amnisos, 47–49, 50

Amorgos, 90

Amos, 87

Antarctica, 103, 122, 132, 136, 137

antelope, 132, 184, 204

Antikythera Mechanism, 149*n*

Antiquity, 42, 154

Apollonius of Rhodes, 171–72

Arabs, 106–9

archaeological excavation, 4–5, 17, 19–21, 26–27, 104, 154, 155–58, 163–269
  absent objects significant in, 73, 170–71
  of Athens, 176–77

archaeological excavation (*cont.*)
  of Crete, 38–39, 40, 41, 47–
    49, 199–201, 223, 225, 240
  dogma vs. new evidence in,
    165–66, 167, 234
  experts needed in, 172, 198,
    249
  funding of, 172, 261, 262,
    266–69, 298
  household items found in, 20,
    169–70, 194
  of Kea, 155–56
  living facilities for, 193, 254
  methods of, 20, 178, 179–85,
    194–95, 198–99, 257, 258,
    261
  of Mochlos, 276–77
  of negative impressions, 169,
    182–83, 195, 264, 265, 266,
    268
  politics of, 188
  protection of, 20–21, 178–81,
    264–69
  publicity of, 220–21
  see also dating; frescoes
archaeologists, 19, 26, 163–67,
    238, 259–60, 262–63
  overattributions to religion by,
    196
  respectable subjects preferred
    by, 220
  self-perpetuating dogma of,
    234, 238, 239
Argo, 53
Argonautica (Apollonius of
    Rhodes), 171–72
Aristarchus, 114

Aristotle, 23, 206
Arizona, 96*n*
art, 181*n*, 198, 268–69
ashfall, volcanic, 65, 74–77
  on Crete, 74, 76, 199, 200
  from Theran eruption, 7, 15,
    20, 43, 73, 74, 75–76, 77,
    88–89, 170, 200, 233, 275
ash layers, 6, 7, 8, 73, 74, 77,
    233, 236, 238–44, 286, 301
Asia, 14, 38, 56, 115, 124, 126,
    127, 132, 146
asteroid impacts, 36, 57*n*, 99–
    100, 134, 295
  Australian, 124
Astounding Days (Clarke), xi–xiii
astronauts, 99–100, 117, 252, 253
  ancient, 274
astronomy, 21, 114, 219–20,
    221–22, 252–53
  time gate of, 110, 115, 117–19,
    122, 123, 129–30, 132, 137,
    141–43
Astypalaea, 90
Athens, 16, 42, 155, 175, 199,
    231, 252, 254
  archaeological excavation of,
    176–77
  frescoes removed to, 198, 250,
    291, 299
Atlantes tribe, 46, 86, 91, 108
Atlantic Ocean, 23, 24, 25, 33–
    37, 44, 46, 54, 95, 130, 131,
    297
  age of rocks in, 35–37
  mid-Atlantic Ridge of, 36–37,
    58, 94

seamounts of, 6, 33–35, 94
widening of, 129, 133, 136,
    164, 165
Atlantis, 4, 16, 19, 22–26, 29–
    47, 111, 114, 149, 267–68,
    286, 294
  expeditions in search of, 23–24
  fresh water on, 17–18
  linguistic derivation of, 45–46
  location of, 23, 24, 25, 32, 33–
    35, 37–47, 154
  Marinatos's theory of, 42–43,
    149, 151–54, 187, 189, 220,
    298, 300
  modern legend of, 23–26
  submergence of, 19, 44–45,
    50, 152
  see also Plato
*Atlantis: The Antediluvian
    World* (Donnelly), xi, 36n
Atlas, 45–47
Atlas Mountains, 37, 46, 86, 108,
    109, 126, 131, 133
Augustus Caesar, 116
Australia, 115, 120–21, 136, 146
  asteroid impact on, 124
Australopithecines, 127, 129, 130,
    132, 145
Azores, 24, 36
Aztecs, 24, 48n, 106

B
Babylon, 44, 207, 273, 294
bacteria, 35, 137, 139
  sulfide-metabolizing, 54, 55,
    138

Ballard, Robert, 6, 33–35, 165–
    66, 297
Barnum, P. T., 297
Batiza, Rodney, 57
batteries, 273–74
bees, fossil, 164–65
Berber tribe, 108
Big Bang, 132, 141–42
Big Dipper, 110, 115, 117–18,
    123, 129, 141
Bitter Lakes, 245–46
blood, 78, 79
Borneo, 120, 121
brain, 145–47, 213
Brazil, 99, 110–11, 137
Britain, 113
  see also England
British Columbia, 137–38
bulls, 16, 40, 41–42, 112, 199,
    201, 231, 242
  at Catal Huyuk, 116
  in cave paintings, 118
Burgess shale fossils, 137–38
Byblos, 78

C
California, 6, 77, 234–35, 236,
    239
Cambrian Period, 137–38, 139
Canaan (Palestine), 22, 81–82,
    86–87, 207, 241, 245
Canada, 35, 137, 138
Cantor, 23
Cape Horn, 105
Caphtor, 46, 82, 87

carbon-14 dating, 73, 118, 222, 233–34, 235, 238, 239

Carter, Howard, 4, 162, 172, 181, 260

Case, Gerard R., 133, 164

Caskey, John, 155–56

Catal Huyuk, 116, 199

cats, 108, 133

  snakes killed by, 80

caves:

  burials in, 119, 120

  paintings in, 118

Cernan, Eugene, 99–100, 117

Charles, Prince, 95n

Cheng, Harrison, 238–39

China, 104, 106, 110

  historical records of, 236–38, 239

choice evil, 290

Christianity, 90, 292n–93n, 294

  Crusades of, 91, 105, 106, 107, 108–9, 178, 290

  Egyptian religion vs., 240n

Ciparis, Auguste, 66, 71, 297

cisterns, 4, 17, 169, 176, 204, 285

civilization, 14, 90–91, 97, 112–15, 268–69, 279–80, 290–95

Clarke, Arthur C., xii–xiii, 6, 112n, 148–50, 156, 167, 285–95, 297, 300

Cleopatra, 109, 113, 147

Cleopatra's Needle, 302

Clerc, Fernand, 67–68, 297, 299

climate, 56–57, 77, 115, 118, 132, 134, 163

colors, of building stone, 16, 150, 151, 184, 197

Columbus, Christopher, 105, 111n

Compère-Léandre, Léon, 66–67, 71–72, 299

continental drift, 56, 100, 102–3, 104, 109, 113, 115, 122, 129, 139

  of Africa, 37, 128, 130, 131–32, 133, 135, 136, 137

core samples:

  deep-sea, 6, 74, 77, 88, 130, 233, 236

  of ice layers, 77, 235, 236, 238

Cortez, Hernando, 48n

Cousteau, Jacques-Yves, 287–88, 290–91, 297

Crapper, Sir Thomas, 204

Cretaceous Period, 132, 133, 135, 164–65

  see also dinosaurs

Crete, 4, 8, 13, 16, 38–50, 79, 88, 95, 99, 112, 117, 152, 153–54, 157, 186, 199–203, 205–6, 274–80

  archaeological excavation of, 38–39, 40, 41, 47–49, 199–201, 223, 225, 240

  ashfall on, 74, 76, 199, 200

  as Caphtor, 46, 82, 87

  devastation on, 48–49, 62, 73–74, 76, 90, 154, 199, 226, 278–79

  frescoes of, 39, 44n, 153, 197, 200, 206, 228, 230

Greek occupation of, 22, 38–
39, 40, 73–74, 76, 85–86,
146, 155–56, 201, 230, 231,
242–43, 277–80, 293
as Keftiu, 43–44, 45, 46–47,
78, 152, 177, 207, 242
population of, 200, 205
Praisians of, 202–3
refugees in, 228–29, 277, 279
see also Minoans; *specific cities*
Critias (Plato), xv, 18, 29–30, 32,
39, 40–42, 45, 146, 151–52,
153, 154, 202, 205, 231
Crusades, 91, 105, 106, 107,
108–9, 178, 290
Cyclades, 38, 85, 88, 94, 99, 106,
154, 155, 186, 192, 201
Cyprus, 75, 76, 77, 200
in Turkish-Cypriot war, 254–
57, 261
Cyril, Archbishop, 107, 198

D
Daedalus, 38, 39, 105, 202, 203
daggers, 24, 209, 280, 281
darkness, 49, 64, 77, 81
Theran eruption as cause of,
13, 43, 77–78, 81, 171–72,
232, 233
Dart, Raymond, 127
Darwin, Charles, 103, 122, 292
dating, 4, 5–6, 115, 154, 217–46
arbitrary, 234
of ash layers, 6, 7, 8, 73, 74,

77, 233, 236, 238–44, 286,
301
carbon-14, 73, 118, 222, 233–
34, 235, 238, 239
Chinese historical records in,
236–38, 239
of Egypt, 78n, 223, 238–45
ice layers in, 77, 235, 236, 238,
239
peat bogs in, 6, 56n–57n, 235,
239
potassium-argon, 35, 100, 239
pottery styles in, 6, 200, 217,
222–30, 234, 236, 238–40,
241–44, 252, 260, 276–77
tree-ring, 6, 77, 118, 234–35,
236, 238, 239, 268
death clouds, volcanic, 49–50,
59–63, 64–65, 67–83, 299
of Theran eruption, 39, 42n,
43, 62, 74, 75–76, 78–83,
89, 233, 274, 281, 282
deep-sea core samples, 6, 74, 77,
88, 130, 233, 236
Deiss, Joseph Jay, 61–62
Delta House, 184–85, 194–99,
210, 218, 219, 257, 261, 262,
263, 266, 281, 282
Democritus, 114
Deucalion, flood of, 245
dinosaurs, 8, 35, 93, 102, 116,
118, 133–34, 135, 136, 145,
171, 253, 259–60
extinction of, 56–57, 124, 134,
164n–65n, 220, 234
Diodorus, 89

dolphin motif, 200, 224, 228, 229
donkeys, 137, 149–50, 157, 158, 167
Donnelly, Ignatius, xi–xii, 36
Doumas, Christos, 7, 17, 19, 20, 42–43, 73, 175–88, 193–99, 204, 236, 245, 287–88, 298
    background of, 175–77, 178
    banishment of, 251, 253–54
    children of, 261–62
    excavations directed by, 258, 260–69, 281
    Marinatos's relationship with, 178, 185–89, 193–94, 221, 249–54
Drake, Sir Francis, 105

E
Egypt, 4, 5, 13, 19, 62, 86, 88, 90, 94, 113, 118, 124, 126, 177, 184, 200, 207, 254, 259, 294, 298–99, 301–2
    ash layer of, 6, 7, 8, 74, 233, 238–44, 301
    Atlantis legend originated by, 22–23, 30–32, 38, 42–45, 152, 154, 177, 231
    dating of, 78n, 223, 238–45
    frescoes of, 44n, 153, 196, 197, 206, 241–44, 277, 279
    Greeks and, 22, 30
    pharaohs of, *see specific pharaohs*
    plagues of, 77–81, 233, 239, 286–87, 289, 298
    Ptolemies of, 109
    pyramids of, 24, 25, 95n–96n, 114–15, 133, 205, 273, 302
    religion of, 45, 196, 240n
    Thera settled by, 7, 8, 20, 91, 101, 109–10, 147–48
Egyptian blue pigment, 240
Egyptologists, 165, 238, 239, 240, 245
Einstein, Albert, 84, 222
elephants, 126, 132, 153, 200, 285
    fossil, 117, 125, 131
Empire State Building, 95
England, 5, 106, 117, 122, 178, 298–99
Etesian winds, 156–57
Ethiopia, 122, 128, 129
Etna, Mount, 236
Euclid, 273, 274
Euripides, 88, 89
Europe, 37, 38, 56, 102, 113n, 115, 117, 118, 120, 123, 128, 131, 132, 135, 146, 154
evolution, 101–2, 112n, 132–36, 137–39, 163–64, 253
    human, 119–30, 132, 133, 145–47, 289–90, 292–93
Exodus, 6, 80, 81–83, 86, 239, 241, 286–87, 298, 301
    route of, 244–46
Exploration of Space, The (Clarke), 149

F
false winters, 77, 235–36
felix culpa (happy fault), 287

Fernando, Mervyn, 6, 286–87, 288–94, 298

Fira, 98, 101, 137, 150, 155, 157, 167, 222, 233–34, 254, 266, 291

foraminifera, 133

forests, 118, 121, 127, 131, 132, 164

fossils, 17, 35, 76, 116, 128, 130, 135, 136

  Burgess shale, 137–38

  elephant, 117, 125, 131

  insect, 118, 164–65

  marine, 7, 8, 35, 95n–96n, 101, 133, 138, 163–64

Fountains of Paradise, The (Clarke), 285

Fox, Sidney, 140

France, 56n, 106, 118, 125, 129

Francheteau, Jean, 53

frescoes, 4, 17, 18, 26, 27, 90, 112, 179, 191, 204–7, 266

  artists of, 196–97, 228, 277

  of Crete, 39, 44n, 153, 197, 200, 206, 228, 230

  of Egypt, 44n, 153, 196, 197, 206, 241–44, 277, 279

  Egyptian blue pigment in, 240

  naval fleet depicted in, 184, 192, 205, 228, 241, 259

  removal of, to Athens, 198, 250, 291, 299

  restoration of, 169, 183–84, 194–98, 258

  women depicted in, 39, 184, 200, 204, 206–7, 227, 249–50

Fresco of the Ladies, 249–50

Fresco of the Lilies, 194–98, 228, 254, 269

Frost, K. T., 39–41, 42, 47, 298, 300

Furumark, Arne, 230

G

Galanopoulos, Professor, 245

geographic barriers, 113–14

geometric motif, 223, 224

Germany, 102, 117, 148, 154

Ghost from the Grand Banks, The (Clarke), xii

Gibralter (Pillars of Hercules), 23, 32, 38, 42, 44, 91, 95, 108, 111, 130, 153, 177

Glomar Challenger, 35–37, 58, 130

Goddard, Robert, 148

gold, 73, 78, 114, 115, 125–26, 141, 171

  in recovered artifacts, 42, 62, 104, 199, 209, 274, 280, 281–82

Gondwanaland, 136

Gould, Stephen Jay, 134–35, 290n

Greece, 38, 90–91, 94, 99, 103, 155, 242, 268

  Ionian, 114

  military government of, 100, 172, 186, 249, 255, 256

  in Turkish-Cypriot war, 254–57, 261

  see also Athens

Greeks, 22–23, 30, 44, 110, 153, 176–77, 199, 229–30, 274
  Crete occupied by, 22, 38–39, 40, 73–74, 76, 85–86, 146, 155–56, 201–2, 230, 231, 242–44, 277–80, 293
  Linear B script and, 201, 202, 230, 240, 278
  mythology of, 31, 38, 88, 151, 156, 253
  Thera settled by, 8, 20, 91, 101
greenhouse effect, 237n, 264
Greenland, 6, 132, 235, 236, 238, 239

H
Halley's comet, 98, 103
Hammer, Claus, 235
hanging gardens, 26, 27, 169, 194, 227
happy fault (*felix culpa*), 287
harbor, Theran, 157, 191–92, 204, 219
Hatshepsut, Queen, 43, 223, 224, 240–41, 244
  curse of, 298–99
Hawking, Stephen, 112n
Hebrews, 22, 81–82, 86, 87, 202n, 244, 245, 287, 298
Herculaneum, 61–65, 69, 74, 110, 171, 178
Herodotus, 46, 153, 201, 202–3
Hitler, Adolf, 102
Hittites, 22
Hollis, Ralph, 54–55
Homer, 153, 171, 192–93, 201

Homo erectus, 124–25, 126, 127, 145, 146
Homo habilus, 127
Homo sapiens, 119–24, 146, 292
House of Sorrows, 27
Hyria, 203

I
Iakovidis, Spyros, 279, 280
Icarus, 38
Ice Ages, 34–35, 56n, 115, 117, 118, 127, 132, 150, 235
  interglacial periods of, 119, 121–23
Iceland, 211, 212, 215
ice layers, 77, 235, 236, 238, 239
Indonesia, 77, 115, 236n
insects, fossil, 118, 164–65
Ionian Greece, 114
Ipuwer papyrus, 43, 78–79, 81
Iran, 14, 116, 119
Ireland, 6, 234, 235, 239
iron, 139
Irwin, James, 252, 253
Islam, 91, 103, 106–9, 294
isolation, effects of, 112n, 115, 127
isotopes, 35, 100, 238
Israel, 22, 82, 86, 120, 246
Italy, 37, 85, 90, 112, 128, 202–3

J
Japan, 106, 116
Java, 49, 121, 127
Jeremiah, 82

Jerusalem, 91, 108, 109
Jesuits, 6, 82–83, 86, 286
jiffy, 143
Johanson, Donald, 129
Julius Caesar, 110, 113, 147

K
Kea, 38, 100, 157
  archaeological excavation of,
    155–56
Keftiu, 43–44, 45, 46–47, 78,
    152, 153, 177, 207, 242
Kennedy, John F., 149
Kenya, 127
Killian, John and Christy, 60, 61,
    69n
Kir, 87
Knossos, 40, 41, 62, 73, 112, 154,
    172, 184, 199, 206, 228, 230,
    240, 242, 244, 278
Kos, 75, 76, 90
Krakatoa, 49, 53, 89
Kubrick, Stanley, 149, 177
Kydonia, 200

L
La Brea tar pits, 265
Labyrinth, 16, 38
Landes, Gaston, 65–66, 299
language, 125, 145, 199
Late Minoan 1A, 1B, (LM1A,
    LM1B) pottery styles, 217,
    222–30, 234, 236, 238–40,
    241–44, 252, 260, 276–77
Laurasia, 136

Leakey, Louis, 127
Libya, 38, 109, 205
lilies, 194, 197, 204, 224, 228
Limbrex, Susan, 260
Linear A script, 199–202, 205,
    230, 236, 278
Linear B script, 201, 202, 230,
    236, 240, 278
Little Ice Age, 56n
Lord, Walter, 147–48, 213, 299–
    300
Lost Atlantis (Luce), xv, 43
Love Canal, 8, 211–13, 215, 300
Luce, J. V., xv, 43, 45–47, 199,
    244, 278–79
Lucy, 129, 130

M
Makarios, Archbishop, 255
Mallowan, Sir Max, 7n
mammals, 135, 253
Marinatos, Nanno, 167–68, 170–
    72, 192, 196, 210, 255–57,
    258–59, 260, 279, 300
Marinatos, Spyridon, 6–7, 42–
    44, 46–49, 76, 91, 101, 109–
    10, 150–58, 201, 249–59,
    260, 265, 266, 268
  archaeological excavation by,
    168–72, 178–89, 191–99,
    258
  astronomy loved by, 219, 221,
    252–53
  death of, 100, 194, 257, 258
  Doumas's relationship with,

Marinatos, Spyridon (*cont.*)
    178, 185–89, 193–94, 221,
    249–54
  grave of, 194, 262–63
  health of, 251
  political views of, 186, 249,
    256
  ships loved by, 192–94
  temperament of, 185–86, 199,
    249–51, 253, 254, 256
  theories of, 42–43, 149, 151–
    54, 187, 189, 217–29, 232,
    234, 236, 252, 298, 300
marine fossils, 7, 8, 35, 95*n*–96*n*,
    101, 133, 138, 163–64
Mars, 25, 118, 177, 252, 302
Mavrorachidi Ridge, 185, 192
Mayans, 24, 114, 211
Mediterranean Sea, 13, 16, 23,
    33, 39, 57, 86, 103, 133,
    135, 136, 152
  evaporation of, 130–31
Melos, 13, 233
Mesa Vouno, 7, 101, 107–8,
    109–10, 147, 168, 192
Mesozoic Era, 135
Meteor Crater, 96*n*
Mexico, 24, 25, 36, 48*n*, 115, 211
Michelangelo, 105
mid-Atlantic Ridge, 36–37, 58,
    94
Minoans, 4, 16, 19, 22, 38–49,
    85–86, 110, 126, 134, 153,
    185, 205–7, 238, 240, 253,
    294–95
  astronomy and, 219–20, 221
  bulls worshipped by, 16, 40,
    41–42, 112, 116, 199, 201,
    231, 242
  clothing of, 39, 44*n*, 153, 184,
    200, 204, 206, 241–42, 244,
    249, 277
  culture of, 24, 26, 39–42, 45,
    112
  decline of, 38–39, 42, 47–49,
    76, 154, 202, 207, 263–64,
    274–80
  diet of, 170, 179
  earthquake-resistant buildings
    of, 281
  empire of, 38, 39–40, 111–14
  naval fleet of, *see* naval fleet,
    Minoan
  printing by, 200
  royal monuments lacking
    among, 206
  scripts of, 199–202, 205, 230,
    236, 240, 278
  size of, 195
  slaves of, 206
  women, 39, 156, 184, 200, 204,
    206–7, 227, 249–50
  see also Crete; Thera
Minos, King, 16, 31, 49, 202,
    203, 253
Minotaur, 16, 31, 38, 41
Miocene Epoch, 130, 132, 163
Mochlos, 276–77
Monolithos, 156
moon, 13, 19, 99–100, 114, 118,
    177, 252, 253, 292, 302
  Taurus-Littrow Valley of, 117,
    123, 126, 133, 136, 139
Morocco, 45–46, 130, 133, 135

Moses, 6, 44, 81, 240, 244–45, 287, 289, 302
Mouttet, Louis, 65, 299, 300
Munier, Al, 100
Murray, Douglass, 298–99
Mycenae, 277, 279, 293

N
NASA, 100–101, 149, 237
nasal morphology, 122–23
Natural History, 290n
naval fleet, Minoan, 16, 48, 202–3, 227, 278
    depiction of, in frescoes, 184, 192, 205, 228, 241, 259
    destruction of, 20, 42–43, 274–76, 279
Naxos, 4, 13, 38, 106
Neanderthal man, 119–21, 122–23, 124, 146, 197
    burials of, 119–20, 147
negative impressions, 169, 182–83, 195, 268
    resilient materials needed for, 264, 265, 266
Nelson, Bruce, 60, 61
Neolithic Age, 86
nesting behavior, 209–15, 300
    panic vs., 213–14
New Jersey, 133, 164–65
New York, 24–25, 277, 299, 302
    continental drift and, 102, 109, 129
    Love Canal in, 8, 211–13, 215, 300

New Zealand, 7, 115, 134, 163–64, 236n, 259, 285
Nile River, 6, 8, 74, 77, 79, 103, 118, 130–31, 239, 240, 301
North America, 37, 56, 102, 135

O
oceanography, 22, 53–55, 58, 166, 287–88, 297
    of Atlantic Ocean, 24, 25, 33–37
    deep-sea core samples and, 6, 74, 77, 88, 130, 233, 236
    of seamounts, 33–35
oceans and seas, 15, 16
    age of rocks in, 35–37
    levels of, 56, 57, 117, 119, 122
    see also Aegean Sea; Atlantic Ocean; Mediterranean Sea; Pacific Ocean
octopus motif, 224, 229, 276
Old Testament, 22, 44, 46, 52, 79, 87, 101, 116, 202, 231, 272
    see also Exodus
Oligocene Epoch, 132
ostrakon, 176–77
oxygen crisis, 139

P
Pacific Ocean, 57, 115–16, 120, 135, 297
Paean (Pindar), 156
paleontologists, 26, 163–67, 259

Palestine (Canaan), 22, 81–82,
  86–87, 207, 241, 245
Pang, Kevin, 237–38
Pangaea, 136
Papadopoulos, Georgios, 100,
  186, 255, 256
peat bogs, 6, 56n–57n, 235, 239
Pelean phase, 63, 64, 66, 282
Pelée, Mount, 8, 63, 65–73, 74,
  79–80, 224, 297, 299, 300
Pellegrino, Gloria, xv, 58, 261,
  262, 285
Pericles of Athens, 176–77
Permian Period, 136
Phaeacia, 153
Phaedrus (Plato), 32
Phaëthon, 30
Phaistos, 62, 200, 206
Philistines, 22, 46, 82, 86, 87,
  231
pillars, sacred, 41–42, 45–46
Pillars of Hercules (Gibraltar),
  23, 32, 33, 38, 42, 44, 91,
  95, 108, 111, 130, 153, 177
Pindar, 156
Planck Barrier, 142
plants, 17, 116, 130, 138
  flowering, 164–65
  hybridized, 197
  seeds of, 260
platinum group elements, 134n
Plato, xi, 4, 16–18, 22–23, 26,
  29–32, 33, 37, 38, 40–45,
  46, 77, 111, 114, 151–54,
  184, 201, 230–32, 291
  Critias of, xv, 18, 29–30, 32,

39, 40–42, 45, 146, 151–52,
  153, 154, 202, 205, 231
  Timaeus of, xv, 12, 19, 31–32,
  39, 50, 154, 202, 231–32
Pleistocene Period, *see* Ice Ages
Plinian phase, 63–64, 81
Pliny the Younger, 63–65, 214
Pliocene Epoch, 57, 130
plumbing, 3–4, 17, 18, 24, 25,
  27, 204–5, 207, 274
pollen samples, 56n–57n
Polynesians, 106, 115
Pompeii, 53, 61, 65, 69, 169, 171,
  194
  modern damage to, 178–79
population, 86, 97, 99, 102, 103,
  105, 109, 115, 117, 118, 124
  of Crete, 200, 205
  of Thera, 101, 106
Poseidon Adventure, 275
potassium-argon dating, 35, 100,
  239
pottery styles, 6, 200, 217, 222–
  30, 234, 236, 238–40, 241–
  44, 252, 260, 276–77
Praisians, 202–3
Ptolemies, 109
pumice, 20, 85, 150
  cult of, 225
  mining of, 98–99, 103–4, 155,
  185, 280–81
  rafts of, 76, 254
pyramids, 24, 25, 114–15, 205,
  273, 302
  marine fossils in, 95n–96n,
  133

R

Rakham, Professor, 275–76
Ramses VI, 181*n*
randomness, 294–95
Red Sea, 37, 103, 128
    "Reed Sea" vs., 245–46
Rekhmire, tomb of, 44*n*, 242–44,
    277, 279
religion, 26, 45, 196–97, 259,
    293, 294
    see also specific religions
rhytons, 44*n*, 242
Rio de Janeiro, 110
Romans, 94, 107, 108, 110, 113,
    116, 201, 264, 274, 278, 294
    Americas contacted by, 110–
    11, 114
    Thera settled by, 7, 8, 20, 91,
    101, 147–48
    see also Herculaneum; Pompeii
Roraima, 67, 68–71, 301
Ruff, Sue, 60, 61
Russell, Edith, 208, 210–15, 300–
    301

S

Sagan, Carl, 142, 248
St. Helens, Mount, 8, 49–50, 53,
    58–61, 62, 69, 74, 75, 79,
    80–81, 224–25, 236, 272,
    301
St. Pierre, Martinique, 63, 65–
    73, 74, 79–80, 224, 230,
    281, 297, 299, 300
Samos, 89

San Marino, 96*n*
Sargon of Akkad tablet, 46*n*
Saudi Arabia, 37, 128
Schachermeyer, F., 244
Schmitt, Harrison, 99–100
Schumaker, Larry, 33–34
Scotland, 117
Scott, Dave, 252, 253
Scott, Ellery, 67, 68–71, 301
seamounts, 6, 33–35, 36, 37, 94,
    297
    oceanic life sustained by, 53–
    55
seas, *see* oceans and seas
seeds, 260
Seismosaurus, 133–34
Semut, tomb of, 241, 242
sewage systems, 17, 204, 207,
    211
Shanidar cave, 119
shipwrecks, Roman, 110–11,
    114
Sicily, 99, 108, 202, 203
Sirbonis Lake, 245–46
Six-Day War, 246
skeletons, 104, 180, 185, 197,
    209, 210, 280, 281, 282
Skhul cave, 120
Smayda, Ted, 97
Soles, Jeffrey, 276–78
Solon, 30–32, 38, 42–45, 46–47,
    70, 151, 152–53
Songs of Distant Earth, The
    (Clarke), xii–xiii
South America, 102, 110–11,
    114, 135, 136

space exploration, xi, 19, 98, 99–
101, 148–49, 177, 221–22,
252–53, 291–92, 293, 302
see also moon
Spain, 23–24, 29, 48n, 91, 106,
107, 108, 109, 112, 118,
124–25, 130, 135
squatters, 217–19, 225, 282n
Sri Lanka, 6, 149, 285–86, 297,
298
stadia, 154
Stanley, Daniel, 238–40, 301
"Star, The" (Clarke), 286
stone tools, 128–29
storage jars (*pithoi*), 170, 172,
179, 180, 184–85, 205, 210,
217, 218
stromatolites, 139
Suez Canal, 103–4, 281
Sumatra, 236n
sun, 56n, 114, 118, 302
swallows, 194, 195–96, 204, 224,
228, 229
Syria, 14, 22, 86, 87, 200, 205,
206

T
Tacitus, Caecilius, 63
Taieb, Maurice, 129
Tambora, Mount, 74–75, 77, 79,
235
Tanzania, 127
Taylor, S. R., 275
Teilhard de Chardin, Pierre, 86
Telchines Road, 26, 185, 210,

217, 218, 257, 261, 262, 263,
266
Thera, 3–8, 13–27, 55,
147–58
agriculture of, 104–5
ancient landscape of, 191–92,
204, 205
ancient names of, 15, 16, 20,
83, 89–90, 199, 282
descriptions of, 15, 16, 17, 19,
112, 135, 167–68
fresh water on, 3, 17–18, 26,
27, 104, 168–69, 204–5
geology of, 150, 197, 295
harbor of, 157, 191–92, 204,
219
Lagoon of, 7, 19–20, 57–58,
101, 104, 150, 152, 155, 167,
287–88, 290
Minoan cities of, 3–5, 7, 8,
16–22, 26–27, 38, 89, 91,
96, 103–4, 147–48, 154,
156–57;
see also archaeological
excavation
modern earthquakes on, 101,
168
population of, 101, 106
skeletons and, 104, 180, 185,
197, 209, 210, 280, 281, 282
successive settlements of, 7, 8,
20, 85, 90, 91, 101, 109–10,
147–48
time gate of, 93–143
see also Atlantis; Crete;
Minoans
Thera, eruptions of, 13–14, 15–

16, 20, 22, 42–43, 47–49,
85–91, 117, 150, 154, 197,
232–33, 263–64
in ancient literature, 77–83,
87, 88–89, 171–72, 202, 239
ashfall from, 7, 15, 20, 43, 73,
74, 75–76, 77, 88–89, 170,
200, 233, 275
darkness caused by, 13, 43, 77–
78, 81, 171–72, 232, 233
dating and, *see* dating
death cloud of, 39, 42n, 43, 62,
74, 75–76, 78–83, 89, 233,
274, 281, 282
as divine punishment, 230–31,
291
ejecta volume of, 62
force of, 15, 73, 89
inhabitants evacuated previous
to, 73, 170–71, 184, 209,
229, 279, 280–82
modern, 106, 155
nesting behavior and, 209–15,
300
pumice and, *see* pumice
sequence of events in, 224–26,
280–82
theological aspects of, 6, 286–
95
tsunamis caused by, 16, 22,
42n, 43, 50, 76, 82, 87–88,
89, 90, 203, 254
undersea crater from, 287–88,
290
widespread turmoil caused by,
22, 86–87, 90–91
Theseus, 38, 39

tidal waves, *see* tsunamis
Timaeus (Plato), xv, 12, 19, 31–
32, 39, 50, 154, 202, 231–32
time gate, 93–143, 145–47, 197,
216, 289–90
"Tires and Sandals" (Gould),
290n
Titanic, 6, 27, 55, 70n–71n, 166,
289, 294, 299
Edith Russell and, 208, 210–
15, 300–301
Toba, Mount, 236n
toilets, flush, 3, 17, 24, 204, 218
Tonga, 76
tourists, 150, 168, 184, 220, 265,
267, 300
Towers, Frank "Lucks," 294
toxic wastes, 211–13, 215,
300
tree ring dating, 6, 77, 118, 234–
35, 236, 238, 239, 268
Truman, Harry, 53, 58–60,
301
tsunamis (tidal waves), 47, 48–
49, 75, 156
behavior of, 274–75
Red Sea and, 245
Theran eruption as cause of,
16, 22, 42n, 43, 50, 76, 82,
87–88, 89, 90, 203, 254
Tunisia, 86, 99
Turkey, 13, 22, 62, 75, 77, 87–
88, 90, 94, 103, 104, 106,
126, 127
Catal Hayuk in, 116, 199
in Turkish-Cypriot war, 254–
57, 261

Tutankhamen, 4, 162, 167, 181–
82, 198, 241, 260, 269, 301
Tuthmosis III, 43, 44*n*, 86, 223,
240–45, 273, 277, 287, 298,
301–2
Twain, Mark, 103
2001: A Space Odyssey, 149, 177

U
uranium, 126, 139
User Amon, tomb of, 242

V
Valkyrie, 6
vases, 20, 169, 185, 199, 228, 229
Vatican, 6, 105*n*, 286
human evolution and, 292*n*–
93*n*
Venezuela, 111
Venice, 105–6, 108
Ventris, Michael, 201
vermin, plagues of, 79–80
Vesuvius, Mount, 61–62, 63–65,
171, 178, 214
violence, 201–2, 289–90,
293
Vitaliano, D. B., 274–75
volcanoes, 8, 53–83, 125, 136,
140, 211, 224–25, 236
ashfall from, 65, 74–77
climate affected by, 56–57,
77
darkness caused by, 49, 64, 77,
81

death clouds of, 49–50, 59–63,
64–65, 67–83, 299
largest known explosions of,
236*n*
Pelean phase of, 63, 64, 66,
282
Plinian phase of, 63–64,
81
see also pumice; seamounts;
specific volcanoes
von Braun, Wernher, 148–50,
156, 167, 300, 302
Voyager 2 spacecraft, 221–22,
293, 302

W
Washington, George, 56*n*, 71*n*,
105
water, fresh, 3, 17–18, 26, 27,
104, 168–69,
204–5
Wegener, Alfred, 102–3
Wesson, Robert G.,
113*n*
West House, 157, 183, 184, 191,
192, 194, 204–5, 218, 221,
228, 241, 242, 259, 262, 263,
266
Wheeler, John, 142
winds, 261
Etesian, 156–57
westerly, 6, 77, 233,
275
wines, 105, 252
women:
depiction of, in frescoes, 39,

184, 200, 204, 206–7, 227, 249–50
statuettes of, 156
treatment of, 206
Woods Hole Oceanographic Institute, 33, 53, 58
world mind, 293
World War II, 36n, 148, 154–55

X
Xesté 3 building, 249

Y
Yorktown syndrome, 213n

Z
Zakros, 62, 73, 199–200, 206

ABOUT THE AUTHOR

DR. CHARLES PELLEGRINO wears many hats. He has been known to work simultaneously in crustaceology, paleontology, preliminary design of advanced rocket systems and marine archaeology. He has been described by Stephen Jay Gould as a space scientist who occasionally looks down and by Arthur C. Clarke as the world's first astropaleontologist. He is, with the late Senator Spark Matsunaga, framer of the 1992 International Space Year. At Brookhaven National Laboratory he and Dr. James Powell coordinate brainstorming sessions on the next seventy years; projects currently under design by Powell and Pellegrino range from a global system of high-speed Maglev trains (New York to Sydney in five hours) to relativistic flight (*Valkyrie* rockets) and the conversion of America to solar power.

In the late 1970s, Dr. Pellegrino and Dr. Jesse A. Stoff produced the original models that predicted the discovery of oceans inside certain moons of Jupiter and Saturn. While looking at the requirements for robot exploration of those new oceans, Pellegrino sailed with Dr. Robert Ballard, worked with the deep-sea robot *Argo*, and traced the *Titanic* debris field backward in time to reconstruct the liner's last three minutes. He has since, with James Powell, developed an economically viable means of raising and displaying the *Titanic*'s four-hundred-foot-long bow section.

Through his work on ancient DNA (including a recipe involving dinosaur cells that may be preserved in mouth parts and in the stomachs of ninety-five-million-year-old amberized flies), Pellegrino hopes one day to redefine extinction. His hope—and his recipe—are the basis for the Michael Crichton novel/Steven Spielberg film *Jurassic Park*.